DISCARDED

A Japanese View of Nature

Although *Seibutsu no Sekai* (*The World of Living Things*), the seminal 1941 work of Kinji Imanishi, had an enormous impact in Japan, both on scholars and on the general public, very little is known about it in the English-speaking world.

This book makes the complete text available in English for the first time and provides an extensive introduction and notes to set the work in context. Imanishi's work, based on a wide knowledge of science and the natural world, puts forward a distinctive view of nature and how it should be studied. Ecologist, anthropologist, and founder of primatology in Japan, Imanishi's first book is a philosophical biology that informs many of his later ideas on species society, species recognition, culture in the animal world, cooperation and habitat segregation in nature, the "life" of nonliving things and the relationships between organisms and their environments.

Imanishi's work is of particular interest for contemporary discussions of units and levels of selection in evolutionary biology and philosophy, and as a background to the development of some contributions to ecology, primatology and human social evolution theory in Japan. Imanishi's views are extremely interesting because he formulated an approach to viewing nature that challenged the usual international ideas of the time, and that foreshadows approaches to study of the biosphere that have currency today.

Japan Anthropology Workshop Series (JAWS)
Series editor: Joy Hendry
Oxford Brookes University

Editorial board: Pamela J. Asquith, *University of Alberta*; Eyal Ben-Ari, *Hebrew University of Jerusalem*; Hirochika Nakamaki, *National Museum of Ethnology, Osaka*; Wendy Smith, *Monash University*; and Jan van Bremen, *University of Leiden*

A Japanese View of Nature
The world of living things
by Kinji Imanishi
Translated by Pamela J. Asquith, Heita Kawakatsu, Shusuke Yagi and Hiroyuki Takasaki
Edited and introduced by Pamela J. Asquith

A Japanese View of Nature
The World of Living Things

by Kinji Imanishi

Translated by Pamela J. Asquith,
Heita Kawakatsu, Shusuke Yagi
and Hiroyuki Takasaki

Edited and introduced by Pamela J. Asquith

Seibutsu no Sekai first published in 1941 by Kōbundō shobō

English edition first published in 2002
by RoutledgeCurzon
11 New Fetter Lane, London EC4P 4EE

Simultaneously published in the USA and Canada
by RoutledgeCurzon
29 West 35th Street, New York, NY 10001

RoutledgeCurzon is an imprint of the Taylor & Francis Group

Seibutsu no Sekai © 1941 Kinji Imanishi, © 1993, 2002 Bunatarō Imanishi

Translation © 2002 Pamela J. Asquith, Heita Kawakatsu, Shusuke Yagi
and Hiroyuki Takasaki
Introduction and editorial matter © 2002 Pamela J. Asquith

Typeset in Times by
Integra Software Services Pvt. Ltd, Pondicherry, India
Printed and bound in Great Britain by
TJ International Ltd, Padstow, Cornwall

All rights reserved. No part of this book may be reprinted or
reproduced or utilised in any form or by any electronic,
mechanical, or other means, now known or hereafter invented,
including photocopying and recording, or in any
information storage or retrieval system, without permission
in writing from the publishers.

British Library Cataloguing in Publication Data
A catalogue record for this book is available from the British Library

Library of Congress Cataloging in Publication Data
A catalog record for this book has been requested

ISBN 0–7007–1631–9 (hbk)
ISBN 0–7007–1632–7 (pbk)

Cover photo: Grasshopper nymph (sp. unknown) on a leaf, Hokkaido.
Photo by Hirotaka Matsuka, Tokyo

To the memory of Kinji Imanishi (1902–1992),
and of Jun'ichirō Itani (1926–2001)

Contents

List of figures	viii
Note on the translators	ix
Foreword (Hiroyuki Takasaki)	xi
Foreword (Shusuke Yagi)	xiii
Preface to the JAWS RoutledgeCurzon series	xvii
Editor's preface	xix
Acknowledgments	xxiv
Note on Japanese names	xxvii
Introduction	xxix
Seibutsu no Sekai The World of Living Things by Kinji Imanishi	li
Author's preface	liii

1	Similarity and difference	1
2	On structure	9
3	On environment	21
4	On society	33
5	On history	61

List of terms in the original index	87
Bibliography of publications in Western languages by Kinji Imanishi	90
Index	92

Figures

1	Street in the Shimogamo area of Kyoto leading to Imanishi's house	xliii
2	Ground floor door to Imanishi's study where he wrote *The World of Living Things*	xliv
3	The Takano River, a tributary of the Kamo River, Kyoto, east of Imanishi's home	xliv
4	View of the Kamo River, Kyoto, where Imanishi began his studies of mayfly larvae, west of Imanishi's home	xlv
5	View of the Kamo River, Kyoto, showing the swift current, west of Imanishi's home	xlv
6	Mayfly larvae lifeforms studied by Imanishi in the 1930s (original drawings by Yonekichi Makino): (1) *Ephemera lineata*; (2) *Ephemerella basalis*; (3) *Ameletus montanus*; (4) *Cinygma hirasana*; and (5) *Epeorus hiemdlis*	xlvi
7	Mayfly larvae species that segregated into different habitats in response to the river current studied by Imanishi in the 1930s (original drawings by Yonekichi Makino): (1) *Ecdyonurus yoshidae*; (2) *Epeorus latifolium*; (3) *Epeorus curvatulus*; and (4) *Epeorus uenoi*	xlvii
8	Kinji Imanishi in 1937 or 1938 (photo courtesy of Jun'ichirō Itani)	xlviii
9	Kinji Imanishi with Adolph Schultz, University of Zurich, in 1958 (photo courtesy of Jun'ichirō Itani)	xlviii
10	Kinji Imanishi with Sherwood Washburn, University of Chicago in 1958 (photo courtesy of Jun'ichirō Itani)	xlix
11	Kinji Imanishi with Clarence Ray Carpenter and son at Pennsylvania State University in 1958 (photo courtesy of Jun'ichirō Itani)	xlix
12	Kinji Imanishi in 1982 (photo courtesy of Keita Endo, photo by Setsuo Kono)	1

Note on the translators

Pamela J. Asquith is a Professor of Anthropology at the University of Alberta, Canada. She received a BA (Anthropology and Psychology) from York University, Canada and a DPhil (Biological Anthropology) from Oxford University, England. Her research interests are in the anthropology of science, comparative cultures of primatology, modern Japanese views of nature and the archives of Kiniji Imanishi. Hobbies include historical biography and adventuring with her giant schnauzer, (TH) "Huxley".

Heita Kawakatsu is a Professor at the International Research Center for Japanese Studies, Kyoto, Japan. He received a BA and MA (Economics and Economic History) from Waseda University, Japan, and a DPhil (Economic History) from Oxford University, England. His specialization is comparative socio-economic history and his research interests are intra-Asian competition and British imperial history.

Shusuke Yagi is Associate Professor of Japanese and Asian Studies, Furman University, USA. He received a BA from the International Christian University, Tokyo and a PhD (Anthropology) from the University of Washington, USA. His fields of research interest include transdisciplinary studies, modern Japanese literature and popular literature, non-Western epistemology/ontology, and IT application to classroom teaching.

Hiroyuki Takasaki is Associate Professor in the Department of Biosphere–Geosphere System Science at Okayama University of Science, Japan. He received his BSc and DSc (Biological Anthropology) from Kyoto University, Japan. His research areas are biological anthropology and primatology. His hobby is collecting and breeding butterflies and beetles.

Foreword
As if climbing a favorite mountain

This small book is worth reading many times. I read it in a paperback edition for the first time when I was 17 years old. The pages of my first copy became too loose for easy holding after reading it eight times in five years. Since then I have bought two additional copies, and have ceased to count the number of times that I have read it. A copy accompanied me on many fieldwork trips, and still does. Whenever I read it, in particular high above the earth's surface on intercontinental flights, the opening passages impress me anew. It is unbelievable that this book was written before we witnessed our blue planet from space.

Among the works left by Imanishi, this book constitutes just a small portion. It occupies only about one-third of the first volume of his *Collected Works*, which amount to 14 volumes. Ten volumes were published in 1974 while Imanishi was still an active writer, and four supplementary volumes appeared in 1993 following his death in 1992. In other words, this book comprises less than three per cent of all his printed works. However short it may appear, it bears the essence of all of his work; so he placed it first in his *Collected Works*. The opening three paragraphs in his own preface well explain the reason for the enigma of the iridescence of this work. He wrote this as his "self-portrait" to leave behind in case he died, which he thought was fairly likely in the war. Though written by a biologist, this book does not read at all like a book of biology. It is, rather, a book of philosophy written by a naturalist who was a thinker faithful to his own beliefs.

Imanishi was definitely not an unquestioning follower of established doctrines. He started his career in biology as an entomologist. Dissatisfied in entomology, however, he turned to ecology – the economics and sociology of living things. He wrote this book at this stage in his life. After some more work in ecology, he turned to anthropology, a converging point for comparative animal sociology. It was a natural consequence that he became the founding father of primatology in Japan. His view of the world of living things, which is perceived to have an infinitely nested structure, greatly influenced the discovery of social structures in nonhuman primates by Japanese primatologists. Their discovery of cultural behaviors is also traceable to his world view, which encourages anthropomorphism when judged appropriate. Most important of all, to me, the path indicated by this book eventually pulled me into anthropology and primatology. After retirement he returned to evolution. He was an "Imanishian" evolutionist from the very beginning, as is evident in this book. Throughout his life he was an alpinist, and left a legacy of climbs in the glacier-covered Himalayas in Asia and Ruwenzoris in Africa, as well as 1552 peaks in Japan. In short, he was always a pioneer, seeking challenges, and brought extraordinary gifts to his endeavors.

On the Satsuma Peninsula in Kyūshū, southernmost of the four main islands of Japan, there is a dormant volcanic peak named Kaimondake. It is sometimes called "Satsuma-Fuji" because of its resemblance to Mount Fuji. Although not so significant in height at 922 meters, it demands a few hours of hard climbing to reach the summit because the path, which spirals steeply upward, starts near sea level. The climb is demanding, but the view from the mountaintop is magnificent. Sometimes the summit may be covered with cloud, and the view may seem fathomless in the woolly whiteness. To describe the difficulty of the climb and the splendor of the view, the local people say, "If he who has once climbed Kaimondake climbs it once again, he is a fool. But if he who has twice climbed Kaimondake does not climb it once more, he is a fool doublefold." They mean that Kaimondake is such a mountain that one who loves it should climb it again and again.

Although this book is small, the reader will find it difficult to read in some places, but after all he may find it worth reading. It is not a book to read to criticize, however, as it was written as a "self-portrait." If the reader dislikes it, he should simply throw it away. It is a matter of taste. If he likes it, he should just forget what is written in the ornamental extras – the introduction and the translators' forewords – and read the text again and again as if climbing a favorite mountain.

Hiroyuki Takasaki
Okayama, Japan

Foreword
The meaning of the translation of Imanishi's *The World of Living Things*

For a non-Western person, the period from the late 1960s to the end of the 1980s was an exciting time, which gave a glimmer of hope finally to go beyond the Cartesian dichotomy, positivism, Eurocentrism, illusory objectivity and universalism. Postmodernism, feminism, postcolonialism and multiculturalism challenged conventional objectivistic and realist approaches to knowledge, or so it seemed. Several writers[1] gave me the impression that this could herald a new era. Yes, they were still Eurocentric, but nonetheless they were a good beginning, I thought.

When I was first involved in this translation of Imanishi's *The World of Living Things*, I was hopeful for the trends mentioned above, especially for the emergence of a new trend in anthropology. During the late 1980s, Western ethnographers belatedly realized that what they wrote were narrative texts susceptible to the biases of representation, authority and inequality between the observer and the observed. To rectify the situation by presenting both the Self and Other in ongoing dialog,[2] various experimental writings were advocated[3] but seldom practised. In addition, these

---

1 To name but a few, Homi Bhabha (ed.), 1990, *Nation and Narration*. London: Routledge; Michel de Certeau, 1984, *The Practice of Everyday Life*. Berkeley: University of California Press; Michel Foucault, 1980, *Power/Knowledge: Selected Interviews and Other Writings, 1972–1977*, Colin Gordon ed. New York: Pantheon; Stephen Greenblatt, 1990, *Learning to Curse: Essays in Early Modern Culture*. New York: Routledge; Meaghan Morris, 1988, *The Pirate's Fiancée: Feminism, Reading, Postmodernism*. London: Verso; Richard Rorty, 1979, *Philosophy and the Mirror of Nature*. Princeton: Princeton University Press; Edward Said, 1978, *Orientalism*. London: Routledge and Kegan Paul; Gayatri Chakravorty Spivak, 1988, *In Other Worlds*. London: Routledge.
2 Barbara Tedlock, 1991, From participant observation to the observation of participation: The emergence of narrative ethnography. *J. of Anthropological Research 47*: 69–94.
3 James Clifford and George Marcus, 1986, *Writing Culture*. Berkeley: University of California Press; George Marcus and Michael Fisher, 1986, *Anthropology as Cultural Critique*. Chicago: University of Chicago Press; Renato Rosaldo, 1989, *Culture and Truth: The Remaking of Social Analysis*. Boston: Beason Press.

writings lacked true epistemological reflexivity[4] and any realization of the cultural embeddedness of the endeavor itself.[5] Even one of the very best recent attempts does not escape this characterization.[6] Worse still, already there are some advocating a return to universalism.[7] Although Finkielkraut[8] maintains that the anti-colonial and anti-Western sentiments of Third World theorists are derived from very reactionary sources such as German Romanticism, he does not know how difficult it is for them to make Western intellectuals understand non-Western thought.

A similar trend can be detected in the natural sciences, in which scientific endeavor is seen as a historical and cultural product.[9] Many practitioners however, remain quite ignorant of developments in the humanities and their importance. There has also been the inevitable backlash against these incipient openings of the Western mind,[10] perhaps more within the science community than among the general public.[11]

How can one make others recognize the legitimacy of non-Western discourses in an intellectually hostile atmosphere? In my view, there have been two major ways to do so. One way is to demonstrate how academic discursive practices in Western academic cultures have been no more than a deceptive and futile practice to "render the strange

---

4 Pierre Bourdieu and Loic J. D. Wacquant, 1992, *An Invitation to Reflexive Sociology*. Chicago: University of Chicago Press; Jean-Paul Dumont, 1986, Prologue to ethnography or prolegomena to anthropology. *Ethos 14*: 344–367; Andrew Strathern, 1993, *Landmarks: Reflections on Anthropology*. Kent: Kent State University Press.

5 Shusuke Yagi, 1991, Japanese ethnography: Searching for the shadow of the other. Paper presented at the 50th Anniversary Meeting of the Association for Asian Studies, New Orleans, unpubl. ms.

6 Anna Lowenhaupt Tsing, 1993, *In the Realm of the Diamond Queen: Marginality in an Out-of-the-Way Place*. Princeton: Princeton University Press.

7 Adam Kuper, 1994, Culture, identity and the project of a cosmopolitan anthropology. *Man* (N.S.) 29: 537–554; SP Reyna, 1994, Literary anthropology and the case against science. *Man* (N.S.) 29: 555–581.

8 Alain Finkielkraut, 1995, *The Defeat of the Mind*. New York: Columbia University Press.

9 For example, Bryan Appleyard, 1992, *Understanding the Present: Science and the Soul of Modern Man*. New York: Doubleday; Morris Berman, 1989, *Coming to Our Senses: Body and Spirit in the Hidden History of the West*. New York: Bantam; Donna Haraway, 1989, *Primate Visions: Gender, Race, and Nature in the World of Modern Science*. New York: Routledge, Chapman & Hall; Mary Midgley, 1992, *Science as Salvation: A Modern Myth and Its Meaning*. London: Routledge; Laura Nader (ed.), 1996, *Naked Science: Anthropological Inquiry Into Boundaries, Power and Knowledge*. New York: Routledge.

10 For example, Paul Gross and Norman Levitt, 1994, *Higher Superstition: The Academic Left and its Quarrels with Science*. Baltimore: Johns Hopkins University Press.

11 For example, Melvin Konner, 1993, *Medicine at the Crossroads: The Crisis in Health Care*. New York: Pantheon; Bill Moyers, 1993, *Healing and the Mind*. New York: Mainstreet Books/Doubleday.

Foreword   xv

familiar"[12] by reducing heterogeneity and otherness, thus paving the way for the oppression, conquest and control of other cultures.[13] Another way is the so-called indigenization of science, creating a new concept or model that is molded from indigenous experiences situated in a non-Western episteme, and applying it with a new meaning in Western scientific discourse.[14] Hsu[15] and Maruyama[16] are two examples of this indigenization, yet Hsu's *jen* and dyadic relationship model in psychological anthropology, and Maruyama's second cybernetics have been either neglected or misunderstood for many years by their Western colleagues. Others, such as Okonogi's[17] Ajase-complex, which emphasizes the mother–child relationship as an alternative to the Oedipal complex, along with Stanley Kurtz's[18] study of the Durga-complex of Hindu India, should be discussed seriously. Finally, more than half a century after Heisaku Kosawa submitted a paper on the Ajase-complex to Freud in 1932,[19] several works have started to appear in English.[20] Shigeo Miki's[21] unique contribution to anatomy, which shows a striking isomorphism with

---

12 Hayden White, 1978, *Tropics of Discourse: Essays in Cultural Criticism*. Baltimore: Johns Hopkins University Press [orig. 1928].
13 Nobumi Iyanaga, 1987, *Gensō no Tōyō – Orientarizumu no Keifu* [The Illusory Orient: A Genealogy of Orientalism]. Tokyo: Seidosha; VY Mudimbe, 1988, *The Invention of Africa: Gnosis, Philosophy, and the Order of Knowledge*. Bloomington: Indiana University Press; Edward Said, 1978, *Orientalism*. London: Routledge and Kegan Paul.
14 Peter Park, 1988, Toward an emancipatory sociology: Abandoning universalism for true indigenization, *International Sociology 3*: 161–170.
15 Francis L. K. Hsu, 1971, Psychological homeostasis and *jen*: Conceptual tools for advancing psychological anthropology, *American Anthropologist 73*: 23–44.
16 Magoroh Maruyama, 1963, The second cybernetics; Deviation-amplifying mutual causal processes, *American Scientist 51*: 164–179; 250–256.
17 Keigo Okonogi, 1991, *Edipusu to Ajase* [Oedipus and Ajase], Tokyo: Seidosha.
18 Stanley N. Kurtz, 1992, *All the Mothers are One: Hindu India and the Cultural Reshaping of Psychoanalysis*. New York: Columbia University Press.
19 Keigo Okonogi, 1978, The Ajase Complex of the Japanese (1). *Japan Echo, vol. 5, no. 4*: 88–105; Keigo Okonogi, 1979, The Ajase Complex of the Japanese (2). *Japan Echo, vol. 6, no. 1*: 104–118.
20 Alan Roland, 1988, *In Search of Self in India and Japan: Toward a Cross-Cultural Psychology*. Princeton: Princeton University Press; David H. Spain, 1992, Oedipus Rex or Edifice Wrecked? Some Comments on the Universality of Oedipality and on the Cultural Limitations of Freud's Thought. In David H. Spain (ed.) *Psychoanalytic Anthropology after Freud: Essays Marking the Fiftieth Anniversary of Freud's Death*. New York: Psyche Press, 198–224; David H. Spain, 1993, Entertaining (Im)possibilities: Chance and Necessity in the Making of a Psychological Anthropologist. In Marcelo M. Suárez-Orozco and George and Louise Spindler (eds), *The Making of Psychological Anthropology II*. Fort Worth: Harcourt Brace College Publishers, 103–131.
21 Shigeo Miki, 1989, *Seimei-Keitai no Shizenshi* [A Natural History of the Morphology of Life], *Vol. 1*. *Kaibōgakuronshū* [Anatomical Papers]. Tokyo: Ubusuna Shoin.

various aspects of nature, has yet to be introduced to the West. Although difficult, this path nevertheless appears potentially more promising than the first.

This translation of Kinji Imanishi's *The World of Living Things*, as an example of the second way, I hope will stir some interest in the cultural embeddedness of scientific endeavor. Here, at least, cross-fertilization is still possible.

Shusuke Yagi
Greenville, South Carolina

Preface to the JAWS RoutledgeCurzon series

Members of the Japan Anthropology Workshop continually carry out deep and insightful research in Japan, and they meet regularly to present papers about their research and to exchange views on the subjects of their study. The fruits of most of these gatherings have eventually appeared in print in a variety of different forms and formats, and we are proud of our collection. However, it sometimes takes several years for our deliberations to be made widely available, and in a country where change flourishes, this is regrettable. The inauguration of a series devoted specifically to the research of the Japan Anthropology Workshop is a step in the direction of speeding up this process, and we look forward to bringing recent research to the readership as soon as we can after it is presented.

Another aim of the series is to present studies that offer a long-term understanding of aspects of Japanese society and culture to offset the impression of constant change that so tempts the mass media around the world. Living in Japan brings anyone into contact with the fervent mood of change, and former residents from many other countries enjoy reading about their temporary home; but some also seek to penetrate less obvious elements of their temporary life. Anthropologists specialize in digging beneath the surface, in peeling off and examining layers of cultural wrapping, and in gaining an understanding of language and communication that goes beyond formal presentation and informal frolicking. I hope that the series will help to open the eyes of readers from many backgrounds to the work of these diligent "moles" in the social life of Japan.

Of course, no one knows Japan quite as well as a Japanese person does, and I am proud to introduce, as a first book in the series, the translation by a Western anthropologist and her Japanese team of the work of a seminal Japanese anthropologist. Imanishi's work is not well known in the English language, but he had some profound ideas about the place of human beings in the living world that do not always confirm theories to date almost unquestioned in the West. Many Japanese are also unaware of the strength of conviction that ordinary people in other countries have about the way the world came about. If this book makes readers of any background rethink even one or two of their long held assumptions about the way the world has developed, it will have achieved the purpose I envisage for the series.

Joy Hendry

Editor's preface

I first heard of Kinji Imanishi while doing doctoral research at Oxford University during the 1970s. At that time, Kinji Imanishi was of interest to me as founder and a key historical figure in Japanese primatology. By a lucky coincidence, I met fellow doctoral student Heita Kawakatsu at this time. He was very interested in and knowledgeable about the ideas of Imanishi, yet he was an economic historian – a field with which Professor Imanishi has never, as far as I am aware, had anything to do. Kawakatsu told me of the importance of Imanishi's first book in Japanese, *Seibutsu no Sekai* (*The World of Living Things*) which, he said, contained the kernel of many of the ideas upon which Imanishi elaborated throughout his life. He obviously had a high regard for Imanishi,[22] but I had at that time no thought of pursuing a broader study of Imanishi's ideas, nor of going to Japan.

By 1981 some of that had changed. Doctorate in hand, a grant from *Monbushō*, and the cooperation of primatologists in Japan provided me with the opportunity to do an anthropological study of their science. In the autumn of 1981, I still did not consider pursuing Kinji Imanishi's career to any greater extent than any other primatologist's career I had come to study. Yet, soon after an initial meeting with professors from Kyoto University, a plan was made for me to visit the famous Professor. I was thrilled at the prospect of actually meeting this living legend.

Thus, soon after the New Year in 1982, on Imanishi's 80th birthday, I was taken to his home in Shimogamo, Kyoto. New Year is an especially celebratory time in Japan, signifying renewal, a cementing of ongoing

---

22 Kawakatsu later dedicated his widely read book, *Nihon bunmei to kindai Seiyō: "Sakoku" saikō* [Japanese Civilization and the Modern West: Second Thoughts on the "Closed Country"]. Tokyo: Nihon Hōsō Shuppan Kyōkai, 1991, to Imanishi. In the book, he devotes an entire chapter, Imanishi [*shizengaku*] e no chūmoku [Focus on Imanishi's *shizengaku*], pp. 152–182, to Imanishi's thought. His afterword describes his intellectual relation to Imanishi.

friendships, apprenticeships, family ties, and a time to ask formally for favors to continue in the coming year. It is a time when students pay courtesy calls to their teachers, and under that aegis Jun'ichirō Itani, his student Hiroyuki Takasaki and I went to Imanishi's home. Shunzō Kawamura, another pioneer of primatology of Itani's generation, joined us there. The lovely, wood-constructed two-storey home was set in a large treed garden behind a wall. Immediately off the entrance hall was Imanishi's sitting room. This was furnished in Western style with chairs rather than tatami mats. There was a forest of books, lining three walls, and overflowing into small stacks all over the floor. We wound a path through the books to face Professor Imanishi, dressed in Japanese kimono and sitting in a high-backed reading chair. He said it was difficult for him to read now as his eyesight was getting weaker – an unkind twist of fate for such an avid reader.

Even at the age of 80 and for a few years afterwards, long past the time when most would relax and enjoy a well-deserved retirement, Imanishi continued to give thoughtful lectures and to publish, constantly reworking and refining his ideas. I did not hear him give the same talk twice in those initial three years I spent in Japan.

Later, when my Japanese colleagues and I undertook an English translation of *The World of Living Things*, my respect for Imanishi deepened. Although the book could be characterized as a natural philosophy, it is surprisingly widely read by laypersons as well as specialists in Japan. This book provides insights into the basis for the ideas that he later developed on evolution, and much else besides (for which see the Introduction). It is, above all, the work of someone with a wide knowledge of the natural world, and is filled with originality, and, as we can now see, some remarkable foresight into later developments in ecology, biosociology, primatology, and anthropology.[23] The book is thus widely regarded as his most important work for understanding his later writings and as a source of inspiration on many different intellectual fronts.

Imanishi's theory of evolution and his anti-Darwinian views have been copiously published and commented upon in Japan. These were briefly

---

23 See Kiyohito Ikeda and Atuhiro Sibatani, 1995, Kinji Imanishi's biological thought. In David Lambert and Hamish Spencer (eds), *Speciation and the Recognition Concept: Theory and Application*. The Johns Hopkins University Press, 71–89; Yoshiaki Itō, 1991, Development of ecology in Japan, with special reference to the role of Kinji Imanishi. *Ecological Research 6*, 139–155, and Yoshimi Kawade, 1998, Imanishi Kinji's biosociology as a forerunner of the semiosphere concept, *Semiotica 120–3/4*, 273–297, and 1999, Subject-Umwelt-Society: The triad of living beings, *Semiotica 134–1/4*, 815–828.

introduced into English in the mid-1980s.[24] However, and as was evident from the flurry of correspondence that followed on Halstead's article, most Japanese and Western scientists remain entirely unconvinced by them.[25] Thus, although much publicity has been given to Imanishi's anti-Darwinism within and outside of Japan, and although Imanishi himself may have come to consider his anti-selectionism as a fundamental outcome of his view of nature, I think that to focus on it is to miss the point of the man and his real influence.[26]

Imanishi's goal was to understand nature, how things got to be the way they are, and how they fit together in the web of life. The simplest questions about nature are the most difficult to answer and Imanishi tackled them with sincerity and courage as a young man. His great contribution seems to me to be that his ideas and studies of nature inspired people to apply them in original ways in different disciplines, even while disagreeing with one part or another of his views. His life and ideas have also evidently inspired people outside of academia in Japan,[27] and that is a great accolade to a scholar.

The World of Living Things (hereafter referred to as *World*) does not read as a series of finite ideas, but develops in ever-widening circles to make in the end a coherent and subtly changed (for the reader) view of the apparently simplest and commonly held perceptions of the natural world. *World* offers new insights on rereading, and probably the most insights for those with the most experience behind them.

---

24 Beverly Halstead, 1985, Anti-darwinian Theory in Japan. The popularity of Kinji Imanishi's writings in Japan gives an interesting insight into Japanese society, *Nature 317*, 587–589. Halstead became aware of Imanishi's work through an article by A. Sibatani, 1983, The anti-selectionism of Kinji Imanishi and social anti-darwinism in Japan, *J. Social and Biological Structures 6*, 335–343. Also in 1983 Sibatani published a paper Kinji Imanishi and species identity, *Rivista di Biologia 76*, 25–42.

25 A. Sibatani, 1986, *Nature 317*, 587–589; M. Sinclair, 1986, *Nature 320*, 492; CD Millar, NR Phillips & DM Lambert, 1986, *Nature 321*, 475; J. Nakahara, T. Sagawa & T. Fuke, 1986, *Nature 321*, 475; A. Rossiter, 1986, *Nature 322*, 315–316; TD Iles, 1986, *Nature 323*, 576; O. Sakura, T. Sawaguchi, H. I. Kudo & S. Yoshikubo, 1986, *Nature 323*, 586; P. J. Asquith, 1986, *Nature 323*, 675. Halstead responded to this correspondence in B. Halstead, 1987, *Nature 326*, 21.

26 See also Noboru Hokkyo, 1987, Comments on anti-Darwinian theory in Japan: human concerns beyond natural science, *J. Social Biol. Struct. 10*, 377–379.

27 These should not be confused with the use made by conservative elements in Japan of Imanishi's ideas as evidence of a "unique Japanese approach" to science, or with the *nihonjinron* [studies of "Japaneseness"] genre of writers, which are merely political, and not scholarly (cf. Peter Dale, 1986, *The Myth of Japanese Uniqueness*. New York: St. Martin's Press).

Much has been written about the difficulties and pitfalls of translation. Japanese presents perhaps more than many languages. *World*, besides, is rather obtuse, so much so that Jun'ichirō Itani once remarked to me that an English translation of the book may be easier for *Japanese* to understand than the original! This is because a translation removes some ambiguities simply because English forces us to do so. This in turn means that our interpretation will be the views to which readers are introduced. In the final editing, I have done my utmost not to interpret Imanishi's statements unduly. While some of the real poetry of the original was beyond my capacities to render into English, some of the prolixities have been shortened and the translation, I believe and hope, retains the original route of Imanishi's intellectual journey with, as he says, its zigzagging path through the complexities of the natural world.

A word on our translation method would be appropriate. In 1987, Kawakatsu and I spent two weeks in Oxford roughing out chapters one through three. He first read passages aloud in Japanese (thus providing the choice of "reading" for the *kanji* or Chinese characters in which the book is written), and then rendered a very free translation in English on tape, while we both looked at the original. The next year the work continued with Yagi, who was a postdoctoral scholar at the University of Alberta. Yagi and I worked for four weeks in the same manner to complete chapters four and five. This initial roughing in of English was quickly accomplished due to their good abilities in English.

The text, which I then transcribed, however, made very little sense in English. Those with knowledge of literal translations will appreciate that they amount to more or less a third language. The roughed-in text, however, provided me with the structure of the sometimes very long sentences, and saved enormous time in providing the character readings for later checking in dictionaries. A further challenge was that some of the Chinese characters used by Imanishi in 1940 have since been dropped from written Japanese and are not found in modern dictionaries. The following summer, in 1989, Takasaki went over the transcription with me in Japan to clarify complete mysteries and point out omissions. I then went back to the original text and embarked on the translation with these considerable aids. At this point, the translation process was greatly slowed down. I found that subtleties of meaning and interpretation were the greatest challenge, and the reader will see that some of the ideas in the text are not very easy to understand in the first place. I was indeed a snail at this work, and my part took five years to complete. Finally, Takasaki kindly reread the final version. I have not agreed with my co-translators on all points, and faults with interpretation should be laid at my door.

Several Japanese warned how very difficult a translation of this particular book would be. Imanishi himself once commented that if any of his work was translated, he would prefer all of it to be translated at once so that his thinking would be better understood. As his collected works run to 13 volumes,[28] this is unlikely to be done. My understanding of the text increased steadily with the reading and rereading, yet I continue to see different things when I read parts of it again. Toward the end of his life, Imanishi did not object to just the one book being translated.

With his intellectual and personal powers so vigorous in his twilight years when I first met him, I can only imagine what an extraordinary star Imanishi must have been in his prime. It was evident that he was a natural leader, an individual who inspired deep devotion and loyalty. Because of his strong character and effectiveness he doubtless also inspired strong disagreement, which in itself can be fertile ground for new ideas. I think that he and his students must have had a very interesting time. It is our hope that the translation will provide Western readers with the opportunity to follow the early intellectual path of a Japanese scientist. Besides their historical interest, such translations are of a special relevance in the growing circles of philosophical, historical and anthropological interest in alternative epistemologies and science.

Pamela J. Asquith
Edmonton, Canada
June 2000

---

28 The fourteenth volume is a list of his publications, mountaineering and other activities (Kinji Imanishi, 1974–1975, 1993, *Imanishi Kinji Zenshū* [The Collected Works of Imanishi Kinji] *vol. I–XIV*, Tokyo: Kodansha).

Acknowledgments

Work on the actual text that has culminated in this translation began about fifteen years ago, but the groundwork for it was set a few years before then. During that time the project has been assisted, directly and indirectly, through the generosity and interest of many kind people.

I am grateful to Vernon Reynolds of the University of Oxford who initiated the contact with Yukimaru Sugiyama of Kyoto University in 1980 to facilitate my study of Japanese primatologists under a *Monbushō* Fellowship. Provision for me to work at the Primate Research Institute (PRI) in Aichi Prefecture was generously granted by Masao Kawai, then Director of the Institute. I am most grateful to Professor Sugiyama, who in turn became Director of the PRI, for his patience and assistance over many years, and particularly for his insistence that I learn and conduct my studies in Japanese right from the beginning.

I am deeply indebted to Jun'ichirō Itani, formerly of Kyoto University, who has been a mentor to me and to this project in every possible way since 1982. His guidance, generosity, sound judgment and intimate knowledge of the background, acolytes, and ideas of his teacher Kinji Imanishi have been indispensable to the project. While this book was at the Press we learned the tragic news of Professor Itani's death on 19 August 2001. His passing is a deep personal and academic loss to all who have had the good fortune to know him.

It goes without saying that the book owes its existence to the efforts of my co-translators, Heita Kawakatsu, Shusuke Yagi and Hiroyuki Takasaki. Kawakatsu first introduced me to some of the Japanese writings of Imanishi in the 1970s before we ever thought about this project. Throughout the years he and his wife Kimi extended wonderful hospitality and stimulating conversations about many things. Shusuke Yagi helped to keep the project on track by giving up some of his postdoctoral research time when at the University of Alberta, and spending a month on the final two chapters with me. His help and interest were indispensable and greatly appreciated.

Hiroyuki Takasaki has perhaps had the most of anyone to do with this and other projects in which I have participated in Japan, and my debt to him is very great indeed. I thank him for the intelligence, wit and rigor he has brought to our work, and his willingness to tackle any and all problems and questions.

Others who have helped considerably through their discussions, writings and feedback at various stages during the work are Yoshiaki Itō, Atuhiro Sibatani, Tadao Umesao, Michael Huffman, Suehisa Kuroda and members of the seminar for nature-study (*shizengaku no kai*), which was held at Kyoto University with Kinji Imanishi over several months in 1982–1983.

At different times Japanese colleagues in primate centers in Japan have unfailingly provided a welcoming and stimulating environment for this and other projects. Though too many to name individually, I wish to acknowledge with deep gratitude the faculty, graduate students and administrative staff at the Primate Research Institute in Inuyama, Aichi Prefecture; the Institute for Human Evolution Studies and the Center for African Area Studies of Kyoto University, and the Department of Comparative and Developmental Psychology, Osaka University. Additionally, conversations with Motoo Tasumi, then of the Department of Zoology, Kyoto University, as well as writings and conversations with Mariko Hiraiwa-Hasegawa of Senshu University have been helpful in gaining other perspectives on Imanishi and his influence.

I thank Bunatarō Imanishi and his wife Kazuko who graciously allowed me to see the rooms where his father read and worked on his book. I am further indebted to them for their help and cooperation in enabling publication of this translation. Imanishi's second son, Hidejirō Uji (who passed away in 1999), and his family kindly introduced me to many sights in Osaka when I first arrived in Japan. And I thank Kinji Imanishi for including me in several meetings where he spoke, as well as for his active participation in the *shizengaku* seminars held during 1982–1983 at Kyoto University.

Thoughtful responses to the completed manuscript have added to my perspective of how Imanishi's work may be understood outside of Japan. For these I thank Thomas Hoover, Alison Jolly, Gary Prideaux, Vernon Reynolds, Thelma Rowell and David Shaner.

Of a nonacademic but equally important nature was the kindness and generosity of Akira and Masuyo Murase and their two children who became my home-stay family in Osaka shortly after my arrival in Japan in 1981. Over those first six months they introduced me to the culture, cuisine and language of Japan, and much more, the warmth and kindness of a Japanese family. It has been an enormous pleasure to know them throughout these years, even if the passage of time has been more insistently marked by their then 12 and 10 years old son and daughter now having married and begun families of their own. I thank too Keiko Nakatsuka and Suehisa Kuroda of Kyoto for their friendship over these many years.

I make grateful acknowledgment to the Japanese Ministry of Education, Science and Culture (*Monbukagakushō*, formerly *Monbushō*) for funds for the initial work in Japan. I am also indebted to the Izaak Walton Killam Fund at the University of Alberta for funds to meet with Kawakatsu in Oxford in 1987 and for time for writing. I am grateful to The Japan Foundation for a Fellowship to work on the translation with Hiroyuki Takasaki in Japan for two months in 1989.

The drawings of mayfly larvae that appear in Figures 6 and 7 have been reprinted in several sources, including the collected works of Imanishi, and on flyleaves of books about Imanishi. I am grateful to Kazumi Tanida and Hiroyuki Takasaki for tracking down the original Japanese source in a paper published by Imanishi in 1940. The illustrator was the late Yonekichi Makino. Photos of Kinji Imanishi that appear in Figures 8–11 were kindly provided by Jun'ichirō Itani. Permission to reprint the splendid portrait of Imanishi at 80 years of age that appears in Figure 12 was provided courtesy of Keita Endo, owner of *Ryutsu Keizai Shinbunsha*, and taken by Setsuo Kono. To all these people I make grateful acknowledgment.

Finally, a special thanks to my husband, Gary Prideaux, for sharing a delight in Japan, and for his support and forbearance, as he has lived with "the Imanishi project" for as long as he has known me.

I am very glad to have had the opportunity to know a little the person whose thought has occupied mine for several years. I am grateful to all those who made that possible and who have helped in so many ways to make the translation a reality.

Note on Japanese names

Japanese names appear in Western name order, first name followed by family name. This is a departure from the style that many Japanologists have adopted in their publications in English, in which they have followed Japanese name order when referring to Japanese authors. I was emboldened to make this departure because it was Imanishi's own usage in English, and he is better known to English language readers this way. Additionally, Western name order for Japanese active after the Meiji Restoration (post-1868) has been formally adopted by the Japanese editors of *The Japan Foundation Newsletter*, which reports on research on Japan, and is distributed throughout the world in English. They note that many Japanese prefer seeing their names in Western order when in a Western context (Vol. 25/ No. 1, June 1997).

Romanization of the names of modern Japanese writers is not always consistent in the published literature. In each case, the most frequently used, or the most recent form is written here (e.g. Atuhiro Sibatani, not Atsuhiro Shibatani). Similarly, romanization of Japanese book and journal titles in this work is in the form in which they appear in current publications.

Introduction

IMANISHI'S BACKGROUND

Imanishi was born in Kyoto on January 6, 1902, first son, though not firstborn, of a silk textile manufacturer in the area of the city known as *Nishijin*. He thus had the benefit and freedom that a certain wealth and privilege provide. For 500 years, the name *Nishijin* has been synonymous with the production of fine silk textiles. This northwest corner of Kyoto was once the locus of *oribe*, weaving communities attached to the Heian court of noblemen and women of 1000 years ago. Later, textile-producing guilds connected to great families made this area their home.[29] Even today some 20 000 weavers' looms continue to produce silk for the silken kimono, but it past its peak of operations in the late seventeenth–early eighteenth centuries.

Although Imanishi was the eldest son, he did not continue in the family tradition of employment, but in 1932 he moved to a home in the pretty Shimogamo area of northern Kyoto which was close to the Kamo River (Figures 1 and 2). Dyers of kimono material used to dip their cloth further downstream in the fast-flowing waters of this shallow river. Imanishi instead followed his early interest in studies of nature and many of his later scientific studies were carried out close to home in the torrents of the Kamo River (Figures 3–5). Like many children, Imanishi was fond of collecting insects in primary school, but he displayed an early interest in what were to become two lifelong passions – natural history and mountain climbing. For Imanishi, mountains equalled nature as they were one of the few places in Japan free from intensive cultivation or urbanization. When he entered Kyoto University, he chose the College of Agriculture rather than the

---

29 Gary P. Leupp, 1992, Overview of a *Nishijin* Ward in the Late Tokugawa Period, *The Japan Foundation Newsletter*, vol. XX, no. 2: 14–17 [see pg. 14].

xxx *Introduction*

College of Science because he wanted his summers free for mountain climbing. When he married, he named his first and second sons, Bunatarō and Hidejirō, after mountain peaks around Kyoto and Mino, near Osaka. His first daughter, Minako, was named after the highest peak in Kyoto Prefecture, and his second daughter, Madoko, after the mountaineering term for a pass.[30]

In 1928, Imanishi received his Bachelor's degree from the College of Agriculture, Kyoto University, specializing in entomology. His first papers written between 1930 and 1940[31] were based on his studies of the ecology and taxonomy of mayfly larvae (*kagerō yōchū*) of various genera (*Ecdyonurus, Epeorus, Ameletus, Ephemera, Cinygma, Baetis, Paraleptophlebia, Ephemerella, Baetiella, Siphlonurus, Potamanthodes, Bleptus* and *Heptagenia* spp.) in Japanese rivers (Figures 6 and 7). In 1940, he received his Doctor of Science degree from the College of Science, Kyoto University, based on these papers.

During the next decade Imanishi remained around Kyoto University, and was self-supporting, without a permanent position on staff. He is nevertheless remembered for his wonderful seminars, going as often as possible on mountain climbing expeditions which doubled as scientific expeditions. In 1931, Imanishi had founded the Academic Alpine Club of Kyoto (AACK). In many ways his interests in ecology and mountaineering formed the basis for everything else. In the mountains, Imanishi came to regard the study of *living* nature, as opposed to laboratory study of confined or dead specimens, as of paramount importance.[32] Concerned that he would be drafted to fight in the war and might not survive, he wrote his main ideas in *The World of Living Things*, completing it in November, 1940 (Figure 8). He wrote spontaneously and quickly, relating the views that had supported

---

30 J. Itani, personal communication. *Mado* is a colloquial term for a passable path and *ko* is a suffix designating "child".

31 These include a series of ten papers, titled "Mayflies from Japanese Torrents I [through] X". No. I was published in the *Taiwan Hakubutsugaku Kaihō* [Taiwan (Formosa) Natural History Bulletin]; nos. II–IX in the *Annotations Zool. Japon* and no. X in the *Memoirs of the College of Science, Kyoto Imperial University, Series B*.

32 Years later he related that the sight of a grasshopper on a leaf suddenly made him realize that the *real* life of animals is their life in their natural surroundings, not as a specimen in a laboratory or collector's box. He resolved then and there to abandon his collecting and to study the life or "living" of animals (K. Imanishi, 1957), *Reichōrui kenkyū group no tachiba [The standpoint of the Primates Research Group], Shizen* [Nature], vol. 12, no. 2: 1–9.

his biological work thus far, and out of which he developed most of his future ideas and projects.[33]

During the war, Imanishi was sent to Mongolia, which was comparatively safe. With his pre-war experience of four trips to inner Mongolia and one to northeast China for research and exploration, this was a natural choice. However, his scientific collaborator, Tōkichi Kani (1908–1944) was sent to the Pacific and died within the year.[34] Among Imanishi's various trips, he went in 1942 with students Jiro Kawakita, Tadao Umesao, Sasuke Nakao and Tatsuo Kira to the northern part of Da Xinggan Ling in China. In 1944 Imanishi became Director of the Japanese *Seihoku Kenkyūsho* (Northwest Research Institute) in Chōkako in China.[35] He left Mongolia in 1946 when the Institute was closed in the aftermath of the war. Shortly after his return to Japan, he initiated various naturalistic animal behavior studies, which soon became focused on Japanese macaques. Japanese primatology was founded through Imanishi and his students' efforts. In 1950, at age 48, he became a lecturer in the Institute for Humanistic Studies at Kyoto University, though he is best remembered by his former students sitting on tatami in a ground floor room in what used to be a parasitology laboratory annex to the Department of Zoology. Sadly for the nostalgic among us, the building has been demolished.

During the 1950s Imanishi made further mountaineering expeditions to the Himalayas. On a six-month trip through Africa, Europe and America in 1958, he and Jun'ichirō Itani looked for possible African field sites for future primate studies, and visited several pioneers of the developing primate study centers in the West (Figures 9–11). Soon after this, Imanishi directed his students to begin preliminary surveys of the elusive great ape species (including gorillas and chimpanzees) in the Congo, Uganda and Cameroon, as well as, in 1962, to study hunter-gatherers and nomadic pastoralists in Tanzania. In 1959, he became Professor of Social Anthropology

---

33 Syunpei Ueyama, former professor of philosophy at Kyoto University, wrote in his epilog to the 1972 reprint of *World* that he judged Imanishi to be one of the few original thinkers in Japan since the Meiji Restoration (1868). He said that although Imanishi's book is about natural history, it is really a philosophical work and that no previous book had attempted to discuss the basis of biology in Japan.
34 See T. Kani, 1981. Stream classification in "ecology of torrent-inhabiting insects" [1944]: An abridged translation, *Physiol. Ecol. Japan 18*, 113–118.
35 Eiichi Kasuya provides names of people with whom Imanishi worked, many of whom were academic colleagues and students: E. Kasuya, 1993, *Senchūki no Chūgoku ni okeru nihonjin chishikijintachi no kurosurōdo: Chūgoku de no Imanishi Kinji o megutte* [Japanese intellectuals' crossroads in wartime China: Kinji Imanishi in China]. *Gendai Shisō [Contemporary Thought] 21*, 226–231.

at Kyoto University. Astonishingly, from a modern perspective, Imanishi also developed the Laboratory of Physical Anthropology and in 1962 was appointed Professor there too. After his retirement from Kyoto University in 1965, he held the position of Professor of Cultural Anthropology, College of General Education, Okayama University, and in 1967 he became president of Gifu University.

Imanishi's extraordinarily broad and pioneering scholarly career was matched only by his mountaineering career and exploration (on which he also wrote prolifically). The latter was very much the basis for his inspirational example and popularity among the general public in Japan. His personal qualities and contributions were recognized twice by the Japanese government: in 1962 he received the award "Person of Cultural Merit" and in 1979 he was named to "The Second Order of the Sacred Treasure." Imanishi remained active and prolific in publishing and contributing to academic meetings well into his eighties (Figure 12) and only since about 1986, when his sight began to deteriorate seriously, did he curtail his public and scientific engagements. He went blind in 1988 and passed away on June 15, 1992. Fifteen hundred people attended his funeral, including an envoy from the Emperor.

IMANISHI'S INTELLECTUAL CONTEXT AND CONTRIBUTION

Imanishi's life spanned four eras in modern Japanese history. Born in the last decade of the Meiji era (1868–1912) his education had a mixture of traditional and modern or international influences. The Meiji era had heralded an extraordinary change of course for Japan from a nonindustrial feudalistic country to a significant international power. By 1905, at the end of the Russo–Japanese War, Japan had become the first Asian nation to defeat a European nation at war, ironically speeding Japan's acceptance into the Western sphere of influence.

Not many years before, until 1854, Japan had been to all intents and purposes a country closed to foreigners. During the preceding Tokugawa era (1615–1867) only a small amount of exchange occurred between Japan and a colony of Dutch, who were allowed to remain on a man-made island off the port of Nagasaki in western Kyushu, as well as with China and Korea. In 1720 the blanket censorship on all foreign books was relaxed and certain scientific works made their way into the country via the Dutch colony. By the early eighteenth century, Western science was seen largely as a means of providing military strength in the form of guns and battleships. Much later, as the Opium War of 1840 in China proved with alarming conclusiveness, Japanese military technology would be no match against Western weapons. The opening of the country in 1854 was accomplished largely on a protectionist argument. Japan needed Western knowledge to defend itself against such incursions in future.

What became most important in the debates about whether or not to adopt the foreign technology was a problem of ethics and values. The most common charge leveled against the Western spirit and scheme of values was preoccupation with profit to the neglect of duty. Putting one's own pleasure and advantage above dutiful consideration of one's place in family and society flew in the face of the ideals of an ordered Confucian (Tokugawa) society.

Yet others realized that to produce her own science Japan needed to understand the ideas that had led to the appearance of this technology in the West. However, the thinking that had made possible the rise of science – the particular view of the external world and man's[36] relation to it – was not at all compatible with the views on these subjects that had been accepted since the beginning of the seventeenth century in Japan. Thus, Japan needed to rethink some of her most fundamental and sacrosanct assumptions about the way the universe worked. As an example, in Confucian ideology, nature appeared as a vast moral organism, ordered on principles which at the same time ordered the ideal workings of the mind of man and the manner in which one should live in society. The passage of the seasons and the movements of the stars, the way a hawk flies and a fish leaps, were manifestations of the same ultimate principles as those which prescribed that men should be filial to their parents and loyal to their lords.

Soon after the Meiji Restoration of 1868, a group of scholars attempted to solve this profound dilemma in what has become known as the Japanese Enlightenment.[37] They tried to create a new basic discipline in what should be considered valid and useful knowledge, a new scheme of values, a new practical morality in everyday life, a new view of the past and a new theory of political obligation.

The nucleus was a group of scholars who formed the *Meirokusha* society, named for its formation in the sixth year of the Meiji era (1873). Yukichi Fukuzawa (born 1835) was the most expressive exponent of the enlightenment doctrines and wrote copiously for the general public as well as for the government. The idea of importing Western technology while retaining Eastern ethics was, he argued, futile. He pointed out that "civilization" was

---

36 The term "man" is retained here in keeping with the nomenclature of the time.
37 Carmen Blacker, 1964, *The Japanese Enlightenment: A Study of the Writings of Fukuzawa Yukichi*. Cambridge, Eng.: Cambridge University Press (see especially chapter three, "The Enlightenment").

not a matter of "things" but of the way people thought. The "spirit" of the West, he felt, was characterized chiefly by "independence." Because the Western nations had cultivated a spirit of independence, initiative and responsibility, he averred, they had been able to develop their sciences and become strong and prosperous. Lack of such a spirit had made the Japanese fall behind – and this deficiency was to be blamed on Chinese learning and the feudal system of which it was the philosophical justification.

Enlightenment thinkers had, therefore, to prove that the differentiating quality of civilized peoples – the national spirit of independence which produced science and a wealthy nation – would also promote the moral destiny of man. The theory of progress told them that the essence of civilization was the free exercise of independent reason, inborn in all men, not something attained by a genius peculiar to the Western peoples. However, already by 1881 the government was sending visiting Western scholars home and returning to the canons of the Confucian ethic – a policy which culminated in the Imperial Rescript on Education of 1890 that effectively perpetuated the old Confucian virtues for several more decades. The most outspoken of the enlightenment scholars, Fukuzawa, died in 1901, a year before Imanishi was born.

A second point historically relevant to Imanishi's intellectual development is his relationship to the Kyoto School of philosophy and particularly the work of the philosopher Kitarō Nishida (1870–1945). From the middle of the nineteenth century, Japanese thinkers had been interested in Western philosophy. Because they believed that this form of thought had no parallel in their tradition they developed a new word to translate the idea of Western philosophy, *tetsugaku*. This created some distance between traditional forms of Japanese thought such as Buddhist and Confucian doctrines and Western modes of analysis, speculation and argumentation.

The Kyoto school of philosophy[38] had its roots at Kyoto University among specialists in the philosophy of religion. It is considered to have its foundation in Nishida, whose writings mainly span the period from 1910–1945. Imanishi used to visit Nishida's household in the late 1920s and 1930s. One of Imanishi's closest student protégés, the primatologist Jun'ichirō Itani (b. 1926), relates that one of his earliest memories was falling into a garden pond at Nishida's home when Itani was four years old. Itani's father, an artist, was also a regular visitor to the Nishida household. The ties among them go back a long way.

---

38 See Frederick Franck (ed.), 1982, *The Buddha Eye: An Anthology of the Kyoto School*. New York: Crossroad; Thomas P. Kasulis, 1982, The Kyoto School and the West: Review and Evaluation, *The Eastern Buddhist, 15(2)*, 125–144.

The motivation behind Nishida's writings,[39] as with his Meiji predecessors, was the problem of how Western science and Eastern morality could coexist within a consistent philosophical system. Nishida hoped to reveal the universal source of both empiricism and religious/ethical/aesthetic intuitionalism. I do not know how profound was Imanishi's understanding of Nishida's or any other philosophy. However, a modern philosopher, Syunpei Ueyama, has pointed out that whole passages from Nishida's first book *Zen no Kenkyū* (*An Inquiry into the Good*) can be found in Imanishi's *The World of Living Things*.[40] Indeed, several of Nishida's statements in *An Inquiry into the Good* (especially Chapters 10–12, "The Sole Reality," "The Development of Reality through Differentiation" and "Nature") reveal many resemblances to Imanishi's arguments as set out in his own Chapter 1 ("Similarity and Difference") in *World*. Nishida argued that reality is a unified whole, but as such it must include opposition. He wrote: "In such mutual opposition, the two entities are not totally independent realities, for they must be unified; they must be part of the development of one reality through differentiation."[41] As will be seen, the perspective that everything developed and differentiated from one thing is absolutely fundamental to Imanishi's views on the relatedness of all things, living and nonliving, in the world.

Consistent with his Buddhist background, Nishida was suspicious of the substantialization of the self. Even the Kantian or Husserlian transcendental ego seemed to Nishida an abstraction not directly derived from concrete experience. Rather, he argued, the self of self-consciousness is not an entity at all, but an act. It is an acting-intuiting. This concept is left obscure, but the main point seems to be that judgment is possible only as an interactive flow from the person into the world (as the attitude, the action, the intentionality) simultaneous with the flow of the world into the person (as the givenness, the sensation, the presence). Thus, at the basis of every judgment is the interaction of the person with the world and the world with the person. The two cannot ultimately be separated. Subjectivity and objectivity are two profiles of the same event. For Imanishi, the subject (living thing) and the environment were part of each other, flowed into each other, and created a particular world over which each living thing had some control.

---

39 See Kitarō Nishida, 1921, *An Inquiry into the Good*, transl. by Masao Abe and Christopher Ives, 1990, Yale University Press. Also see Keiji Nishitani, 1985, *Nishida Kitarō*, transl. by Seisaku Yamamoto and James W. Heisig, 1991, University of California Press.
40 Stated at the *shizengaku* seminars, 1983, Kyoto University.
41 Pg. 64, Abe and Ives, 1990, ibid.

Nishida also expressed antipathy toward the trend to mechanistic explanations of nature with which same antipathy Imanishi concluded his career as a scientist. In his chapter on "Nature" Nishida noted:

> The present tendency of science is to strive to become as objective as possible. As a result, psychological phenomena are explained physiologically, physiological phenomena chemically, chemical phenomena physically, and physical phenomena mechanically.[42]

Nishida further expressed the view:

> The various forms, variations, and motions a plant or animal expresses are not mere unions or mechanical movements of insignificant matter; because each has an inseparable relationship to the whole, each should be regarded as a developmental expression of one unifying self. For example, the paws, legs, nose, mouth, and other parts of an animal all have a close relation to the goal of survival, and we cannot understand their significance if we consider them apart from this fact. In explaining the phenomena of plants and animals, we must posit the unifying power of nature. Biologists explain all the phenomena of living things in terms of life instincts. This unifying activity is found not only in living things, but is present to some extent even in inorganic crystals, and all minerals have a particular crystalline form. The self of nature, that is, its unifying activity, becomes clearer as we move from inorganic crystals to organisms like plants and animals (with the true self first appearing in spirit).[43]

A further point of comparison is Imanishi's species-society concept for which he coined the term "specia." A species-society is not the equivalent of a biological species but is a sociological concept that implies a system consisting of all members, however dispersed, which support a species through their interactions. The concept of specia is absolutely central to his views of the interconnectedness of things in nature. It is not just a conceptual construct, Imanishi noted; it is an existent entity with an autonomous nature, whose various individuals are continually contributing to the maintenance and perpetuation of the specia to which they belong. It is this expression that is most obviously compared with the thought of Nishida. Nishida argued that an individual could not exist outside the context of society and that a society was the only meaningful,

---

42 Pg. 69, Abe and Ives translation, ibid.
43 Pg. 70, Abe and Ives translation, ibid.

everlasting entity, while individuals were ephemeral. We may note too, that Imanishi did not think that the term individualism (so important in the minds of the Meiji Enlightenment thinkers) in the Western sense should be applied to himself: his view was that the individual can be poetically unified with nature without any self-assertion, though he concedes his scientific career and endeavors would be classed as individualistic by Western thinkers.

Towards the end of his career, Imanishi's writings became explicitly anti-Darwinian,[44] culminating in an expression of "anti-science." He wrote prolifically about this,[45] but only in Japanese, until 1984 when he published a paper in English on the conclusion to his study of *shizengaku* (nature-study).[46] In reflecting on the meaning of his life's work in the article, he repeatedly and almost exclusively returned to ideas he wrote in *World*. In reflecting on his long career, he noted that:

> When I was young, I was engaged in entomology. I have also dabbled somewhat in ecology. At nearly fifty years of age, I switched to the humanities, had contact with the nomads of Mongolia, and in Africa observed gorillas and chimpanzees. After seventy I took up the theory of evolution, and now I am trying to wrap it up. ... I have done quite a few things, but it seems to me that I have been consistent in working on the problem of 'What is nature?' And I feel that it is not the constituent nature as represented by such-and-such-ology, but total nature, that I have been in constant quest of. What I have been seeking all this time is shizengaku.

---

44 Imanishi, however, noted that he was a faithful follower of Darwin in terms of the idea that "somewhere, long ago, the first organism came into being, subsequently dividing and developing, until we arrived at the flourishing nature of the present day" (1984: 362 [see note 46 for ref]). What he opposed was the assumption that organisms had to compete with each other for survival. He explained this in much the same terms as he had used in *World* 40 years before, that as all things had come from the differentiation and development of a single thing, and not as immigrants from other planets, it made no sense to think they would be fighting for space, rather than live as part of a harmonized whole (ibid., p. 365).

45 In books such as *Watashi no shinkaron* [My Theory of Evolution], 1972, Shisakusa; *Dāwinron – Dāochaku shisō kara no rezisutansu* [Dārwinian theory – Resistance from Native Thought], 1977, Chūōkōronsha; *Dāwin o koete-Imanishi shinkaron kogi* [Beyond Darwin – Lecture on Imanishi's Evolutionary Theory], 1978 (with Takaaki Yoshimoto), Asahi shuppansha; *Shizen to Shinka* [Nature and Evolution], 1978, Chikumashōbo; *Shinkaron mo shinka suru* [Evolutionary Theory also Evolves], 1984 (with Atuhiro Sibatani and Shōhei Yonemoto), Libropōto, and in numerous papers and interviews.

46 K. Imanishi, 1984, A proposal for shizengaku: The conclusion to my study of evolutionary theory, *J. of Social and Biological Structures* 7, 357–368.

He continued,

> ...shizengaku does not fit within the general scheme of academic disciplines.... the term 'nature' is being bandied about these days more than ever.... I wonder just how deep an understanding of nature the people who use these words really have. It seems that, though the word nature is used more than ever before, there has never been a time in history when people have had such a small realization of what nature really is.... We must teach (students) that nature is not matter, it is (a) living thing; it is the colossal maternal body, the giant, the behemoth within which we, along with all the other myriad creatures, have always been nourished.[47]

Imanishi concluded that in introducing his ideas to the world, it is not the ecologist Kinji Imanishi, or the anthropologist Kinji Imanishi: it is Kinji Imanishi, the scholar of shizengaku. When he stated that he was no longer a scientist, he meant that he did not agree with current science, which is not addressing the larger picture. Imanishi felt that the ultimate concern and responsibility of a scientist should be to free contemporary people from their cultural fragmentation by making them more conscious of the way art, morality, religion and science have become specialized, censorial, and constrictive to the wholeness of our cultural experience. In *World*, Imanishi has, essentially, written an ethic of how to relate to and understand nature.

The interweavings of modern and traditional nineteenth and early twentieth century thinking continue to provide inspiration to writers today in Japan, as shown cogently in Morris-Suzuki's recent book.[48] In her chapter on "Civilization," the intellectual influences of Tokugawa Japan and of Kitarō Nishida are evident in the writing of three modern scholars whose work she discusses – that of philosopher Syunpei Ueyama, economic historian Heita Kawakatsu and philosopher of science Shuntarō Itō. These writers address the history of civilization and examine Japan's place and role in modern civilization.[49] Here again we may find it useful to weave another thread into the fascinating tapestry of Japanese social intellectual history and contemporary thought through English translation of Imanishi's original work.

---

47 K. Imanishi, 1984, ibid: 366–367.
48 Tessa Morris-Suzuki, 1998, *Re-inventing Japan. Time, Space, Nation*, M. E. Sharpe.
49 The titles of these popular works are: S. Ueyama (ed.) 1990, *Nihon bunmeishi no kōsō* [A Plan of the History of Japanese Civilization] 7 vols. Tokyo: Kadokawa Shoten; H. Kawakatsu, 1991, *Nihon bunmei to kindai Seiyō: "Sakoku" saikō* [Japanese Civilization and the Modern West: Second Thoughts on the "Closed Country"]. Tokyo: Nihon Hōsō Shuppan Kyōkai; S. Itō, 1990, *Hikaku bunmei to Nihon* [Comparative Civilization and Japan]. Tokyo: Chūō Kōronsha [cited in Morris-Suzuki, 1998].

READING THE TEXT

It is the editor's intention as far as possible to allow the reader to derive his own insights and to make her own intellectual links with Imanishi's text. If his work is to be read as a philosophy of biology, this is essential. However, a few guideposts will be helpful to aid readers in the initial stages of his argument. These remarks will be confined to the ideas developed in this book, rather than follow their later trajectories. The surprisingly modern ring of some of his ideas as compared with current commentary in biological science, as well as in studies of the interface between animal behavior and human social evolution studies are evident, but beyond the scope of the introduction. With that in mind, the following comments are intended as signposts rather than a detailed map of his ideas in *The World of Living Things*.

The book contains five chapters, of which the first three are intended as an introduction to the discussion in Chapter 4 ("On Society") and Chapter 5 ("On History"). To Imanishi, Chapter 4 was the heart of the book, and Chapter 5 followed as an extension of his discussion on society.

In the first chapter, "Similarity and Difference," Imanishi refers to "change" of the earth from a single thing to the multiplicity of related forms we now see. He states that change is not mere change, but is a kind of growth or development. What Imanishi meant was that things have a place and a function in the structure of the whole world. This place is not predestined, but the fact of a thing's existence means it is there for a reason – it fits – having developed along with the unfolding of the whole world. "Unfolding" again does not imply preformation, but rather simply that things must fit because they developed from within the fold.

Similarity and difference in the chapter title refer to our recognition of things as they are in their relation to each other. Things of this world have different degrees of resemblance and difference. Our recognition of their relatedness or affinity is the simultaneous perception of similarity and difference among things. We are able to do this because all things developed from one thing, and all things are related both in terms of blood and soil or living space. Herein lies the basis for our intuitive understanding of similarity and difference. The closer the relation, the more empathetically we can understand them. That is because the distance and closeness between things in terms of affinity reflects the distance and closeness of the particular environments or worlds in which they live. It follows, therefore, that though we humans have the world of humans, monkeys have the world of monkeys, amoebae have the world of amoebae, and plants have the world of plants, because of the relatedness, we can say that the world of monkeys is closer to ours than the worlds of amoebae and plants, and that the world of living things is closer to ours than the world of inanimate things. Each species is thus similar and also different to the other, and we have an intuitive, naturally determined, subjective response to degrees of affinity.

The themes in this chapter presage the development of Imanishi's interests in anthropology and primatology. Of particular relevance is his statement of the "objective of biology." Biology is not related to the resources for human life, he says, but provides the path by which we can understand our biological affinity with the living world, and that the roots of our behavior are in the world of living things.

The second chapter is "On Structure." Our world is not a random chaos but is an ordered one with a certain structure. Living things are distinguished from nonliving things by their shape or morphology, but also by their complete structure, including their internal morphology. All living things are composed of cells as the structural unit, and develop from a single cell. But nonliving things also have a form. Although we classify living things by their forms in taxonomy, external morphology alone makes no distinction between living and nonliving things. These are distinguished instead by the fact that only living things are composed of cells. More importantly, they are distinct from dead things because they are "alive." Living animals fly or swim because they are alive. That is what "living" means. When we consider living animals in terms of body or structure, flying or swimming are phenomena that cannot be described well by the concept of structure alone. These should be called functions rather than structure. Only a structure that allows various kinds of functions such as flying or swimming can be the structure of living things.

Thus, in Imanishi's idea, structure of a living thing is nothing but for function and vice versa. As the body of a living thing developed from one cell, and if the inseparability of structure and function is a fundamental principle of the existence of living things, then what has developed from one thing is not only the structure of living things, but also the function. The structure of living things is integrated with their function.

Imanishi further notes that the only world we know is one in which everything exists and perpetually changes. This is a world, then, with space and time, as well as structure and function, as one inseparable set. Inanimate things, as constituents of this world, also have integrated structure and function, and in that sense they may be said to have their "inanimate life."

Chapter 3 "On Environment" builds on the idea that the "life" of a nonliving thing is the action expressed by it directed by the integrating process of the life of living things. Thus, the atom of oxygen itself does not change, but its action is under the control of the living body. Living things cannot exist without environment; their way of being can be comprehended only in a system comprising their environment also, and the environment is a part of the world that has coevolved with them. Recognition of the environment by a living thing is the recognition of what is necessary for its living. As it makes a living in an environment, it makes the environment an extension of itself. In the world that the living thing recognizes and utilizes, it is master, in control not only of its own body, but of its environment also. The integrating nature of a living being which consists in controlling and governing itself and its surrounding world seems to be interpretable as their autonomous subjective character (*shutaisei*). *Shutaisei* is a recurring concept in this work. Any living being must be an autonomous subjective being in this world, as it makes a living in it. The subjectivity is a character endowed on living things from their very beginning on this earth; in it lay the root of what eventually developed into human mind. The life of a living thing consists in assimilating the environment and controlling the world, and that is after all the development of *shutaisei* endowed on it.

These initial discussions set the stage for Chapter 4, "On Society," which Imanishi considered to be the core of the book. Right off, Imanishi tackled a question of interest in science then and now: why do organisms live in proximity other than for reproductive purposes?

He asks, "What is a species?" He notes that an individual sees a conspecific as an extension of its own body. This is a basis for his thinking that nature abhors conflict. Imanishi does recognize competition for the same resources, especially food, and says that various life forms emerged due to this. But this was a division of resource-use and the forms that could utilize them, rather than war over resources. Even interspecifically, competition is futile, as may be seen in the mutualism between parasite and host.

Members of a species gather, not for reproductive purposes, but because they have the same needs. In their common habits they find the most stable, and thus the most secure life. That world is the world of the species, and the life there is the life of the species. This shared life does not imply a conscious and active cooperation; rather, as the result of the interactive influences among individuals of the same species, a kind of continuous equilibrium results. The species society is a real entity in this world, or in other words, the world of species is a social phenomenon.

Imanishi later gave the name "specia" to the species society. As the basis of the formation of species societies, every organism is postulated to have an intrinsic faculty of perceiving the identity of fellow members of the same species. *Shutaisei* is the vital attribute of every living individual and species society. Every living thing is considered to be a subjective autonomous entity that acts on and interacts with other living things and its environment. These living things form a species society, which in turn, in a similar manner, acts on and interacts with other species societies to form the whole living world. All of these together were called the holospecia in later publications.[50]

In the fifth chapter "On history," organic evolution is discussed with emphasis on the sociality of living things. Imanishi denies random mutation and natural selection as the prime movers in speciation. He asks, "Can we think of organisms...passively depending on chance?" He thinks that natural selection does occur, but not as the major driving force of speciation. Thus, he contends by way of illustration, that if Australia rejoined with mammalian communities, the marsupials would probably become extinct. This would be due in part to natural selection – but, from the point of view of the whole society of mammals – it would be the resettling of the whole community into a structural equilibrium.

For the title, and throughout the book, the term "living thing(s)" rather than the more usual term "organism(s)" is usually used to translate *seibutsu*. This usage adheres to Imanishi's reflection in Chapter 2 "On Structure," in which he refers to the literal meaning of the Chinese characters *sei-butsu* as "living things." Imanishi notes that living things are first of all conceived to exist as things, and thus they tend to be regarded as merely a physical existence, with "life" left out. He considers that a natural feature of our way of understanding the world is to perceive it first and foremost as a world of things and that "life" has to be tackled on with difficulty afterward. As Imanishi's intent is to discuss the nature of "life" as a whole, the term "living things" reflects his own starting point.

Finally, a word on reading the text itself. The apparently simple and non-technical style of Imanishi's writing sometimes belies the subtleties of his points. This presented many challenges to translation, not only of style, but of rendering the author's intent. Those unfamiliar with Japanese writing will soon realize that an author often returns to a point from different perspectives, sometimes after quite a long digression. A very experienced translator's description of the process is apt for the present text. John Bester[51] noted:

> Nor is the Japanese feeling for the organization of an argument always the same as ours.... There is the tendency of Japanese logic to effect what I can only describe as a spiraling movement. This is not the same as going around in circles. The reader does not find himself back where he started – there has been progress, though not necessarily in the direction in which he appeared to be traveling at any given moment.

---

50 K. Imanishi, 1984, ibid.
51 N. Hasegawa, 1938, *The Japanese Character*. Tokyo. Trans. J. Bester, London: Ward Lock & Co., 1966: xii–xiii.

Introduction xliii

The river of its argument may flow at a leisurely pace, it may have its whirlpools and backwaters, but the river is still there, and the reader who lets himself be carried along on it is likely to find his outlook on the subject at hand subtly changed, whatever reservations he may have about specific conclusions or the methods by which they are reached.

Such a progression differs from the step-wise progression of argumentation with which English readers are familiar. It is, however, no less valid. It may seem more tentative, even less organized, but it should be appreciated as representing a different culture's written and spoken milieu.

The footnotes in the translated text were not in the original. They have been added to help clarify certain points and terms, but have been kept to a minimum in keeping with the spirit and style of the original.

Figure 1 Street in the Shimogamo area of Kyoto leading to Imanishi's house.

Figure 2 Ground floor door to Imanishi's study where he wrote *The World of Living Things*.

Figure 3 The Takano River, a tributary of the Kamo River, Kyoto, east of Imanishi's home.

Figure 4 View of the Kamo River, Kyoto, where Imanishi began his studies of mayfly larvae, west of Imanishi's home.

Figure 5 View of the Kamo River, Kyoto, showing the swift current, west of Imanishi's home.

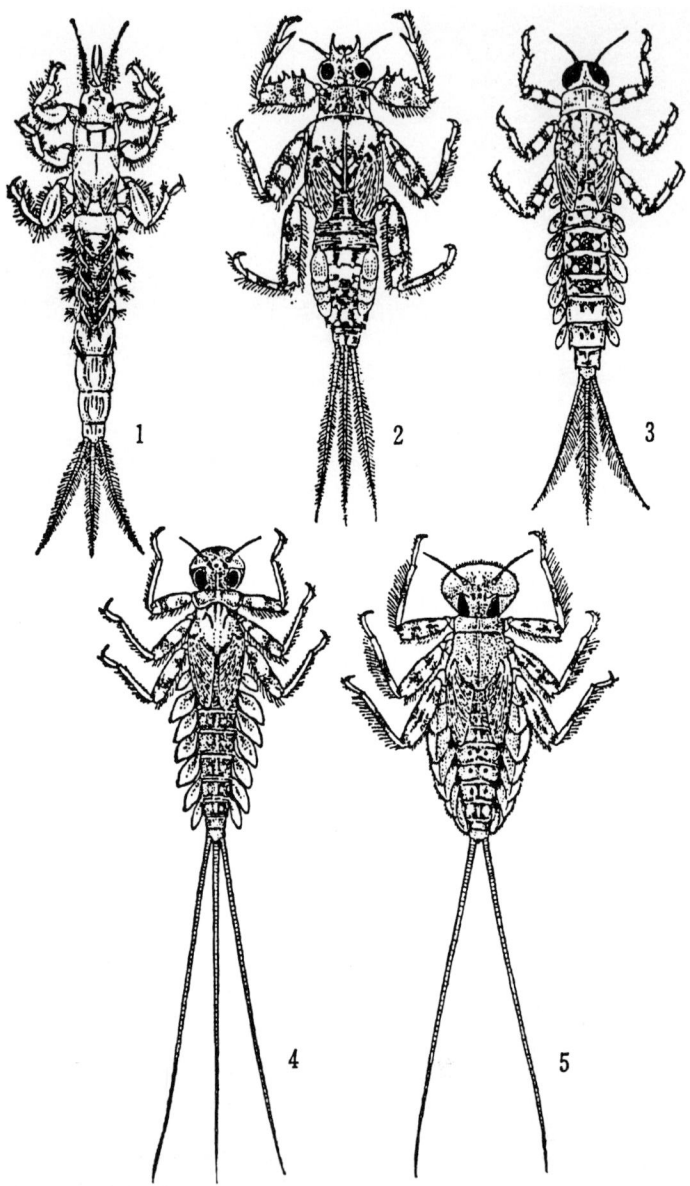

Figure 6 Mayfly larvae lifeforms studied by Imanishi in the 1930s (original drawings by Yonekichi Makino): (1) *Ephemera lineata*; (2) *Ephemerella basalis*; (3) *Ameletus montanus*; (4) *Cinygma hirasana*; and (5) *Epeorus hiemdlis*.

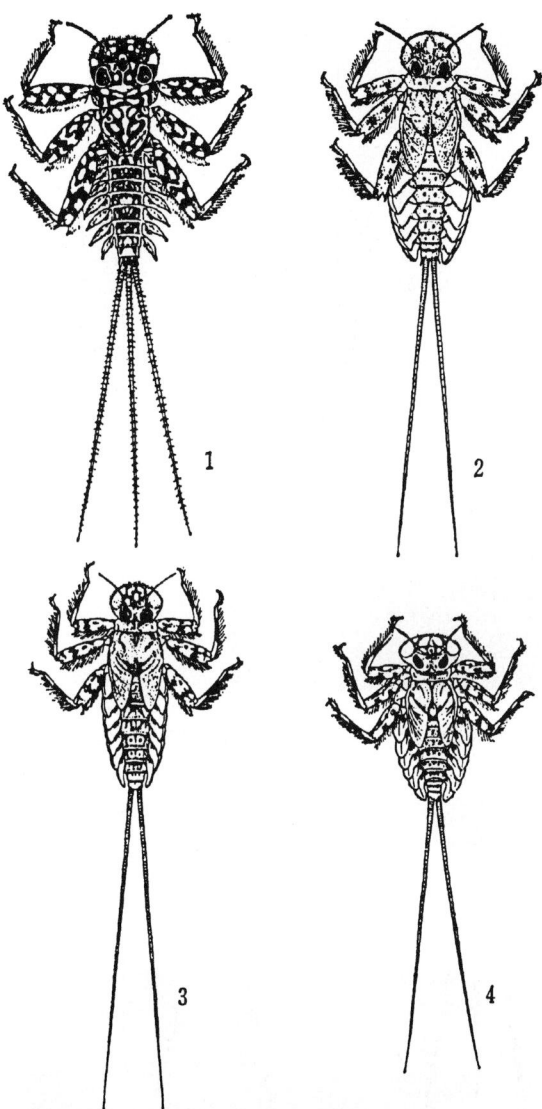

Figure 7 Mayfly larvae species that segregated into different habitats in response to the river current studied by Imanishi in the 1930s (original drawings by Yonekichi Makino): (1) *Ecdyonurus yoshidae*; (2) *Epeorus latifolium*; (3) *Epeorus curvatulus*; and (4) *Epeorus uenoi*.

Figure 8 Kinji Imanishi in 1937 or 1938 (photo courtesy of Jun'ichirō Itani).

Figure 9 Kinji Imanishi with Adolph Schultz, University of Zurich, in 1958 (photo courtesy of Jun'ichirō Itani).

Figure 10 Kinji Imanishi with Sherwood Washburn, University of Chicago in 1958 (photo courtesy of Jun'ichirō Itani).

Figure 11 Kinji Imanishi with Clarence Ray Carpenter and son at Pennsylvania State University in 1958 (photo courtesy of Jun'ichirō Itani).

Figure 12 Kinji Imanishi in 1982 (photo courtesy of Keita Endo, photo by Setsuo Kono).

Seibutsu no Sekai
The World of Living Things

by Kinji Imanishi

Author's preface

What I intend in this small book is not to write a scientific treatise, but to give my personal view of the world, which is the wellspring of my scientific writings. In this sense, this is myself or my self-portrait.

I wished to write a self-portrait because, since the current war[52] broke out, I could have been called upon to serve at any time. I have always liked nature, even as a child, and since graduating from university I have become even more intimate with it through my training in biology. Still, I have not yet left behind any worthwhile contribution, with many plans to do so, and as I may die in this war, my wish is to leave in my very best capacity some record of one biologist in Japan. My fear of running out of time compelled me to complete this work quickly. To accomplish this, no other way was left for me but to draw a self-portrait.

Thus, the book was written very quickly and without a specific plan. Because of the rough style, philosophers or social scientists might disagree with my clumsy use of terms such as cognition or community or culture. My excuse is that I have not yet mastered the art of painting.[53] Where the writing lacks in artistry, if it succeeds in portraying a simple lover of learning accurately, I must be content.

The heart of this book is Chapter 4 on society. It relates directly to my own research on which I have worked diligently for several years. There is so much to say on the subject that often my pen may have slipped over

---

52 The Marco Polo Bridge Incident near Peking, 1937, which marked the beginning of the Sino–Japanese war.
53 Nearly 50 years later, Imanishi described his article on protoidentity as "my second self-portrait." (In Kinji Imanishi, 1987, *Shizengaku no tenkai* [The Development of *shizengaku*] Kodansha, 162–179). The term *shizengaku* has been retained in English language articles. It means nature-study.

things which were not yet clear enough; but if it must be left imperfect, then that must be the self-portrait. In fact, I can claim no chapter to be complete, although I feel confident about some chapters, but not about others. Chapters 1 through 3 are intended as an introduction to the discussion in Chapter 4, but, as I am not a specialist in those fields, I fear that I may not have succeeded in integrating them well with Chapter 4. However, the discussion on history in Chapter 5, which is also my first attempt, I have long wanted to write as an extension to my discussion on society, and it was written at a stroke following Chapter 4. Thus, although the book has its shortcomings, Chapter 4 is the core, with the first three chapters written with less assurance, and the last with some confidence.

While writing this book I was frequently reminded of and reflected on the discussions I have had regularly over the last several years with friends. These centered on ecology, an expanding new frontier in biology. The topics we discussed, such as what direction to take, or how best to tackle the challenges, appear very vague at first. However, for us, many concerns had to do with issues that urgently needed resolving. Like a wanderer in the wilderness, I have zigzagged and wandered in circles through uncharted territory. That I have arrived at all to the stage of putting these thoughts into this book is due to the guidance of many teachers and mentors who have always given me encouragement and support, and to the friends who are constant companions and consultants in the field, sharing all the difficulties. Without them I could not have completed this work. Although I do not mention each person by name, here I would like to express my heartfelt esteem for these good friends.

November, Shōwa 15 (1940)

1 Similarity and difference

Our world consists of an enormous variety of life. We may think of it as a great family of various members. It is no random gathering; each member has a role in constituting, maintaining and developing this family. This is my fundamental view of the world. I did not always hold this view, but arrived at it gradually through years of observing and thinking about what I saw in nature. Since I shall write on the basis of this view I want to make it clear right from the beginning.

I do not see the world as a chaotic or random thing in which members are like chance passengers on a ship, but as having a certain structure or order and each of its members having a function. Although the various things in the world have an independent existence, they are all in fact in some kind of relationship.

Even if passengers have boarded this metaphorical ship independently, the vessel does not accept all comers without limit or regulations. The numbers of first, second and third class passengers have already been decided. There may be some who wished to board first class but could not because they were delayed or all the seats were taken, and who had to be content with second class. Whatever the case, is it likely that these passengers could have come from other worlds?

Of course, not knowing other worlds, I cannot imagine what they would be like. If some passengers did not come from other worlds, the fact they are on board together could be due to only one thing – they were on board from the beginning. In other words, these passengers came into being on the ship. Although they came into being on board naturally, as if they had been normal passengers who bought tickets and took their seats harmoniously, it seems strange that this should be so. We must look for a deeper relationship than the relationship among passengers who bought tickets and came aboard.

This world, the world that I conceive of, is geocentric. Therefore, I would like to confine the world to the earth and continue the metaphor of the earth as a ship. In my metaphor, this large passenger carrier loaded to capacity is compared to the earth, and just as the passengers have not come from another world, so also the materials with which the ship itself is made do not come from outside. It may seem incredible that the earth, originally detached from what is now the sun, which further nourished it with light and warmth, gradually developed into the ship filled with passengers we now see. I shall attempt a credible explanation. First, I would like to consider the change of the earth not as mere change, but as a kind of growth or development. Of course, this is only one of several possible views. I am not trying to persuade those who disagree with this view. Now, during the course of the growth and development of the earth itself, part of the earth became the materials for the ship and eventually took the form of the ship. The remaining parts became the passengers. Thus, the ship did not precede the passengers, nor the passengers the ship. The ship and passengers originally differentiated from a single thing. Moreover, they did not differentiate haphazardly. The ship became a ship in order to take on passengers and the passengers became passengers in order to board the ship. This is a natural conclusion as we cannot conceive of the ship without passengers, nor of passengers without the ship.

Such a clumsy metaphor may be irrelevant but I wish to point out that the various things in this world are in some kind of relationship and the reason for this is that the world has a structure and a function which derive from the growth and differentiation from one thing. This single source is the basis of the fundamental relationship between everything, plants and animals, both living and nonliving.

Still, the relationship among things seems to be very obscure. That is because there are not just one or two, but various kinds of relationships. Here, I would like to consider living things to make these relationships clear, but before doing so I want to address another fundamental question. Previously, I stated that this world is made up of various things; we can say that because we can distinguish among them. However, although I said various things, in fact no two things in this world are exactly the same. Since the space occupied by one thing cannot be shared by any other, it is the occupation of a space that both provides for the existence of a thing and is the root cause of the differentiation of things.

If we are concerned only with differences, we will find everything is different; despite this, is it not wonderful to learn that nothing exists in complete isolation or does not have something similar to it? If things of this world were completely different from one another, there would be no structure, or, if there were, we probably could not perceive it. Moreover, if everything differed from everything else, then difference itself might not be recognizable. Difference presupposes similarity; by recognizing similarities, we can also recognize differences.

Although the things of this world are various, there is nothing with an absolutely independent and unique existence. In this wide world everything has something else similar to itself. But why is this and what does it mean? I will return to this question in detail later, but for the moment I cannot see that the plurality of similar constituents of the world is a result of chance. Thus, it may be possible to say that during the course of the growth and development of this world, the world separated into things which are different but related, and also into things similar to one another with some kinds of relationships.

Thus, similarity and difference existed from the beginning of the differentiation of things from one source. An example of something simultaneously similar and different can readily be seen in offspring and their parents. When we say a child is similar to his parents we can see clearly the likeness, but when we say they differ we can of course find an infinite number of differences. In relationships between such things, similarity and difference are inherently natural. As all the things of this world result from the differentiation from one thing, then as a matter of course this relationship of similarity and difference must hold.

Thus, to recapitulate, I said that we can say that the world consists of various things because we can discriminate among them; but to say discriminate may give the impression that we are concerned only with differences. However, the moment of recognizing something, before consciously distinguishing among things, is not a mechanical, meaningless thing like the reflection of an image in a mirror. I think that we recognize things in relationship to other things in the world. In other words, recognition is the simultaneous perception of similarities and differences among things.

As I am not a philosopher I do not intend to go into epistemology here, but ask the reader to accept my use of the word recognition[54] in relation to the fact that even a small child can in this sense recognize things. You may ask why I use recognition instead of perception,[55] but it is simply my personal taste. However naive this view of the world may be, I think it must be explained consistently in terms of recognition. What I mean by

---

54 *Ninshiki* 認識.

55 *Chikaku* 知覚.

4 Similarity and difference

so-called naive recognition is to grasp the relationship of things intuitively, without consciously comparing them and explaining the basis for discerning their similarities and differences. This intuition is, I would say, an inborn attribute of our recognition. The reason for that lies in our common origin with all other things in this world; we did not suddenly come into being or arrive from another world, but developed in this world so that this inborn recognition is found within us. The various things that make up the world are not entirely different forms; we and they developed together. Our common origin and development bestow a natural predisposition to recognize easily the relationship among things.

I must turn from the abstract concepts of recognition and instinct to consider living things. A discussion of human beings is usually bound to be problematical, whereas discussing other living things is more straightforward and more to my liking. The question is whether my previous points are applicable to all living things. Like us they were born into and developed in this world, and if they can recognize various things in their surroundings in their own way, there can be little difference between them and us in the essential act of recognition. In speaking of instinct in humans I am assuming that we exist basically as one species of living things. I am, in fact, simply considering human phenomena from a biological rather than a humanistic point of view. This neither relegates humans to an animal level nor elevates animals to a human level, but allows us to discuss both on the same basis.

However, living things include both plants and animals. There are both very primitive animals such as amoebae and there are animals closer and more similar to humans, such as monkeys. Animals, plants and humans make up that category I call living things. If I claimed that all the attributes present in humans are found also in other living things and that since humans have consciousness, so also plants are conscious – it would contradict my view that the various things of this world are infinitely different. Indeed, there must be countless differences among things of the world. But, at the same time, because they are not unrelated to each other having grown and developed from one thing, they also must have innumerable similarities.

Here, I would like to introduce the concept of affinity[56] to refer to the relationship of similarity and difference among things in the world. Affinity refers both to the relationship of blood and of soil. It is the relationship of

56 Ruien 類縁.

closeness and distance in historical development and in society. Of course, there are instances of coincidental similarities between completely unrelated things, but generally speaking the affinity relationship presupposes that the greater the affinity between things, the greater the similarities, and the less the affinity, the more differences are evident. I shall explain this in more detail below, but I wish to emphasize here that affinity is based on the fact that the various things of the world originally came from one thing.

Thus, this affinity relationship provides a first criterion for this view of things. Through affinity, things that are similar are closer to each other, and things that are different are distant from each other, so that although everything lives in the same world, the distance and closeness between things in terms of affinity reflects the distance and closeness of the particular habitats in which they live. It follows, therefore, that humans have their own world, monkeys theirs, amoebae and plants each have their own world. Moreover, the world of monkeys is more similar to the world of humans than to the world of amoebae or plants. In more general terms, the world of living things as a whole is more similar to the world of humans than to the world of non-living things. Here is the basis of both our understanding, through our resemblances,[57] of the lives of other things and, at the same time, its limits.

Because the various things in this world are the result of growth and development from one thing, we can recognize this world, and because of their single origin, to recognize those things immediately means to recognize their affinity. As recognition of things means recognition of affinity, so we can make use of the ability to infer similarity. When we talk about inferring similarity we tend to think of it as a kind of thinking, but in my view it is in essence nothing but our subjective response to our recognition of the affinity of all things. This response may be expressed as joy, surprise, fear, love, hatred; they are all our way of reacting to things in this world. Thus, of course our subjective responses vary with the object of our recognition; there is no problem in that, but insofar as recognition is the recognition of the affinity of relationship among things, our subjective response is in fact determined somehow by the affinity relationships themselves. Thus, the limits of our intuitive knowledge of resemblances are also naturally determined.

---

57 Imanishi employed the term *ruisui*, 類推, which is usually translated as "analogy." However, in Imanishi's usage the Chinese characters connote a broader meaning. He may be interpreted to mean by *ruisui* an intuitive logic about similarity and difference, or a degree of empathy with other living things.

The question then must be raised, why does our subjective response vary according to the affinity of relationships among things? Things with a closer affinity live in more similar worlds than those with distant affinities. We can interpret similar worlds in various ways, but they are similar because the subjective observers recognize the worlds in a similar way. That means that if there is a close affinity among things, the subjects' responses to them will be similar. Therefore, if two individuals with a close resemblance meet, the way they recognize each other must be very similar. Furthermore, the responses of the individuals to each other are similar. This mutual recognition leads to a kind of interaction between them. Our subjective response to the recognition of things with close affinity is our reaction to them and I suppose it is a reaction expecting that they recognize us in a similar way.

Where two individuals have a very distant resemblance, however, this mutual reaction does not apply. It is a disappointing reality that a stone or tree will never respond to our greeting but, as I mentioned, our recognition of their degree of affinity with us is somehow intuitive so that even children do not expect a response from stones or trees. However, in the case of animals, particularly advanced ones such as monkeys or dogs, we do expect some kind of reaction from them. Have we not all in fact experienced a response from their side? Though we recognize them as animals, our response to them is similar to our response to other humans. This is because these animals have a closer affinity with us and our response is our expression of that recognition. We, as human beings, convey our recognition with a human kind of expression – we have no other means of expression.

Therefore, in the case of human beings expressing a subjective response of recognition, it is quite natural that we regard the object in human terms and react to it in human terms. It is probably wrong to expect any other reaction from humans. The fact that primitive and uncivilized people had this human-centered view lends support to my view. Herein also is the basis for maintaining that perception of similitude derives from our response as subjects to the recognition of the essential affinity among various things. But still, even among humans it is difficult to fully comprehend all of another person's life because no two things in this world are exactly the same. How much less able are we to interpret the lives of phylogenetically distant animals living in their distinct worlds? This is why comparative psychologists strongly object to anthropomorphic explanations.

Nevertheless, we cannot regard animals as mere automata nor humans as special creations of an omnipotent and omniscient deity. We cannot deny that humans, along with other animals, are the result of the growth and development from one thing, however preeminent humans have become. We can recognize this world because we humans are part of it. Therefore, it is not at all surprising that shamans and poets could talk and listen to nonhuman life such as trees and stones. Still, their nonhuman voice was not the voice of our mouth, nor were they heard with the ear we use. If we can understand this, there is no difficulty in admitting their kind of communicative experience. I do not think that the life that we know exists in the same way in nonliving things. However, there is no problem to admit that nonliving things may have their own kind of life. We can criticize the anthropomorphic view as subjective and unscientific, but similarly we can criticize the view that regards animals as mere matter or automata, despite our subjective response to them as living. The former view regards nonliving things as living things and the latter relegates living things to a nonliving category. Both views are subjective and unscientific. Taken to an extreme, the mechanistic view of living things could be applied to humans too, but this is thought inappropriate and it has been applied only to other animals, probably because we humans do not like it.

Of course, living things cannot exist without a material basis, but they are assuredly different from nonliving things. Yet formerly, so-called natural histories of animals and plants could hardly be distinguished from study of minerals and rocks. This was due to the influence of the time when taxonomy was regarded as biology and studied lifeless specimens. These differ little in material existence from rocks. As humans are living things they are more similar to other living things than to nonliving things, and have a closer affinity to them. Biology must treat living things as living things, not as lifeless specimens. The category of living things includes both animals and plants, advanced and primitive things, and many in between; each inhabits its own world and leads a particular life so that each living thing should be studied in its own proper perspective.

To look at each living thing in its own right actually means making our recognition of them and hence of their affinities more accurate. By doing so, we make our inference of their resemblances more rational and consistent. To reiterate, we as humans attempt to understand the world of other living things from our human point of view. Therefore, we can only interpret and express their world in human terms. Biology that lacks an intuitive knowledge of resemblances can provide only an impoverished, mechanistic view of the living world. We may say that the rationalization of this intuitive understanding of similarity is the essence of the new science of living things.

Thus, when I use expressions such as the society of living things, or the love of living things, or even fine art, which have been regarded as uniquely human, it should not puzzle or disappoint readers. This way of describing things does not imply that other living things are placed on the same level as humans nor that humans are being reduced to the level of other living things. For instance, take the word "society"; humans, animals and plants are different, so their respective societies naturally differ. However, as humans, animals and plants are all living things, and therefore similar in their affinity relationship, so they have an essential similarity in fundamental characteristics. This is not at all surprising, but to admit these characteristics and properly express them in our own words so that we can understand them is the expression of our recognition of these respective living things. To comprehend correctly each living thing in its own place leads to an accurate perception of our human place. The objective of biology is not only related to the resources for our life, but also to provide the materials by which we reflect on our own entity by making it clear that we are part of this living world, having a biological affinity with it, and that the roots of our behavior are in the world of living things in general. I hope that this conveys the reason for discussing similarity and difference in this introduction.

2 On structure

Right at the beginning of this book I wrote that our world consists of various things. Despite this diversity, the world is not chaotic or without order, but has a certain structure. This is not an unconsidered view nor does it stem from a religious feeling; in fact, there is considerable evidence for it. For me, this evidence is found in the living world, as I will make clear below. Since I said that this world has a structure, it is necessary to understand clearly what is meant by structure. Thus, I will commence by returning again to the starting point.

This world consists of various things. In other words, this is a world of things. Because things exist, we can be aware of this world. If there were no things, we probably could not be conscious of our surroundings. Where a thing exists we feel its opposition or meet with some resistance. The fact that even in the dark we can feel the existence of things or that even the blind can perceive the presence of things derives from the fact that things exist in space. However, when we perceive things, we do so mainly with our eyes. So, in most cases we perceive the existence of a thing as a form occupying space. Therefore, we say a thing has a form or that which has a form is a thing. We find it difficult to imagine the existence of something we cannot see or that does not have a form; in fact, until rather recently, people remained unconvinced of the existence of things that did not have some kind of visible form.

If we categorize the various things of our world, we might divide them into nonliving things, living things, and human beings. If human beings are included in the category of living things, then all the various things of this world can be categorized as nonliving or living things. In other words, things that are not living are lifeless things, or, properly speaking, are nonliving things. Thus, to the seemingly obvious question, "What are living and nonliving things?", I had better give a tentative answer before proceeding further. By living things I mean animals and plants. Animals include such things as beasts, birds, fish and insects, and plants include trees, herbs and fungi. These can either provide us with food or be a danger to us; in either case, they have played indispensable roles in our lives and as such our lives have been bound somehow with theirs, for better or worse, for a long time. We are in some sense familiar with them. Thus, from long ago, before we knew about oxygen or hydrogen, we recognized plants and animals to have a similar existence to our own.

However, as I mentioned, we recognize animals as animals and birds as birds because each has its own form. Of course, we do not know animals or birds only by their shape. Animals run and birds fly. When we associate all the habits with the shape, the form of an animal or a bird that we recognize becomes suddenly very clear. Nevertheless, when we recognize a butterfly in an exhibit as a butterfly, or a fish preserved in alcohol as a fish, it is because each has its own specific form. Taxonomists collect dead specimens of living things, classifying them according to similarities and differences in morphology, and attempt to determine the relationships among them. Taxonomic categorization reflects reality to the extent that each living thing does in fact have a certain form in accordance with its relationship to others. In other words, living things have their own forms which reflect their relationships to other living things, and, at least in the natural world, it is impossible for anything other than a living thing to have the form of a living thing. This follows naturally because the world originated and developed from one thing. So, for our ancestors, it sufficed to see and recognize animals and plants with their own eyes. However, discoveries in various corners of the world have revealed the existence of peculiar plants and animals of which our ancestors were unaware. Likewise, we now know of the existence of microscopic things, including microbes, which have a serious impact on our daily lives. This is a very modern achievement in biology, as is the discovery of oxygen and hydrogen. Although we can conceive of a virus so tiny as to be invisible even under the microscope, the foundation of biological science is based on the fact that the various living things of this world are known through their forms. Through this a new concept of living things has been established. Contemporary thinking cannot conceive of living things except as biological living things. More specifically, biological living things are classified in taxonomy based on their morphology. Taxonomists recognize them as living things, assign them species names and register them within scientific circles. However, if this is the case, there is no point in talking about the distinction between living and nonliving things.

Without relying on biologists, but on our own common sense, we rarely mistake living things for nonliving things. That is simply because the things which have the forms of living things are the living things of this world. However, not only living things have their own forms. Nonliving things also have a shape. We can ask if there is any particular trait in the forms of living things that distinguishes them from those of nonliving things. If we look at individual living things, we see that dogs have a dog form and bees have a bee form. However, we cannot imagine the general morphology of all living things from the form of dogs or bees. If we first consider animals as an example, their forms vary greatly. If we attempt to talk about living things in general we must include all animals and also plants. Compare dog, pine tree and amoeba. Can anyone infer from these the general morphology of living things? Then is it a mistake to think we can easily distinguish between living and nonliving things by looking only at their form? I think not. We do indeed distinguish between them by their forms. As I explained in Chapter 1, we can distinguish between living and nonliving things by the inborn powers of recognition that we possess. Although we do not mistake individual living things for nonliving things, still we have no clear criteria by which to distinguish living from nonliving things. We do not recognize dogs and pine trees only in terms of their particular morphological or biological traits. However, if we are asked what is the basis of the distinction in terms of shape between living and nonliving things, we will be lost. This does not mean that we cannot recognize living and nonliving things by their form, but though living things have characteristic forms that differ from those of nonliving things, their appearances are too varied to make the distinction clear in one word. When scientists distinguish between living and nonliving things, they should not be preoccupied about the differences in superficial forms. If we fail to explain our distinction in terms of shape, we must probe into the source of the essential differences.

The form of living things is not restricted to external appearance alone. They are neither hollow, nor like a solid clay object with a uniform interior. This is readily seen in the anatomy of living things that have muscles, veins and other internal organs. These parts are not placed haphazardly but have a certain order and intricate arrangement. If we call the appearance of living things their outer form, we can assume there is an inner form. If we looked at the internal form, the outer form of living things would seem to be determined by the inner form, so as to give it its shape. On the other hand, if we looked at the outer form, the inner form would seem to be determined by the outer form and made to fit into it. Of course, for an individual living thing, neither the outer form nor the inner form precedes the other: both exist from the beginning. They are complimentary and constitute the body of living things. In short, whether we refer to the outer or to the inner form, they constitute one whole of which the body of living things is made. This is not so much form as it is structure. As such, this is the structure of the body of living things.

The statement that living things have their own form can be put in another way, that living things have their own structure. Thus, the distinction between living and nonliving things should not be thought of in terms of appearance, but regarded as a more fundamental thing pertaining to structure. However, since the outer form of a living thing cannot be conceived without its inner form, the structure cannot be understood without considering both outer and inner forms, and the difference in appearance among dog, pine tree and amoeba must reflect the differences in their structure. We have a universal law relating to the structure of living things, based on the discovery of cells: the living body consists of cells. This finding is not only a great contribution of biological science, but is the major foundation of modern biology.

From various points of view, one of the most important characteristics of organisms, whether dog, pine tree or amoeba, is the fact that their bodies consist of an aggregation of cells. Thus, when ascribing the difference between living and nonliving things to differences in structure, we see that in the case of living things the unit of structure is the cell. The cell is the clearest criterion by which living things can be distinguished from nonliving things. So to say that living things have their own structure can be expressed as that living things have cells as units of structure.

Living things are characterized as living by their structure, and their body consists of cells. We can ask why only living things have this sort of structure. If we consider structure alone, then a butterfly exhibited in a box or a fish preserved in alcohol each has its own structure. A butterfly's structure makes it recognizable as a butterfly and a fish with its structure is recognizable as a fish. Even if it is not preserved in alcohol, the fish at the fish market is recognizable as a fish because it has a fish structure. However, these are all dead fish and they are the lifeless bodies of living things. These lifeless bodies have the structure of living things because they were alive once. The body or structure of living things cannot come into being by itself in this world. It is inconceivable that body or structure exists alone in nature without living things. Why is this? Previously, when we considered the form of living things we encountered a similar question. Now, however, knowing that the body of living things consists of cells presents a solution. In multicellular organisms such as ourselves, we can ask from where the numerous cells come that make up the body. They do not come from somewhere else. All these cells originally come from one cell. This may remind readers of the notion I have repeated from the beginning of the book. That is, although this world consists of a variety of things, they originally came from one thing. It interests me greatly that this concept can also be applied to each living thing which itself is a constituent part of the whole world. This is certainly true in the case of living things, as embryology has proved.

Although the cells which constitute the body of living things are of various kinds, they do not aggregate unrelatedly or without order, but are very carefully arranged. In this way an organism's body is formed. In this we see the structure of living things. Such a structure exists as a reflection of the fact that the world has developed from one thing. Living things constitute this world and reflect the condition of the growth and development of the world. Living things must have developed from one cell; this is the only logical way I can think of the world. Thus, there is no structure of living things which was complete from the very beginning. It is the result of gradual development that is the growth of living things. Growth is living; thus, when we talk about the form or structure of organisms, it must be in terms of vital living things. When we talk about organisms as real entities, we must work under the assumption that they are alive.

Indeed, living things are distinct from dead things and to live is meaningful by contrast to being dead. But what does it mean, to be living? When we refer to a fish as being alive we generally think of it as swimming or feeding. A dead fish can neither swim nor feed. However, a dead fish is still a fish and can be recognized as the body of a fish. Thus, for a fish to have lived means that in the carcass something originally lived and operated and so made it alive. Whether it is called "life" or "soul" matters little, but this sort of idea was prevalent for a long time. Although I do not feel it necessary here to go into detail about this senseless dichotomy, if to be alive and to have life are thought to be synonymous, there is no necessary dichotomy. The absurdity and wrongness of this dichotomy lies in the assumption that even life could be conceived to exist as a dual entity, from which arose the idea that the body is ephemeral and perishable while the soul is eternal.

Because the body or structure of living things is inconceivable apart from living things, and as we now know that the whole body or structure of a living thing developed and grew from one cell, there is no room for this dichotomy. Living animals fly or swim because they are alive. That is what "living" means. When we consider living animals in terms of body or structure, flying or swimming are phenomena that cannot be described well by the concept of structure alone. These should be called functions rather than structure. But, of course, function cannot exist without structure. Only a structure that allows various kinds of functions such as flying or swimming can be the structure of living things. With such a structure, living things can exert corresponding functions and can be said to be alive.

As a general way of recognizing the various things which constitute this world, I began with form and so the interpretation of living things also began with form. However, I have proceeded through structure and finally come to function. At this stage, the interpretation of living things becomes more concrete. To summarize again, living things, as they actually live, are neither existence with only structure, nor existence with only function. Structure and function are closely associated; they do not exist separately. Structure is nothing but for function and vice versa; such an existence is called a living creature. I said that the body of a living thing developed from one cell. If the inseparability of structure and function is a fundamental principle of the existence of living things, then what has developed from one thing is not only the structure of living things, but also the function. The original single cell itself must show structurally and functionally an individual existence of a living thing. Each of the various kinds of living cells that are integrated in one individual, and which are derived from one cell, is not only a unit of structure, but must also be a unit of function. When people call a cell an organism, their "organism" does not refer to the living thing that I mean here. Their organism should be reworded as a living body or an organic body in my use of terms. If the cell has such significance for the existence of living things, we can say that the inseparability of structure and function is not only the fundamental principle for living things, but we can extend the application and say that this inseparability is the fundamental principle for the living body.

Because cells have these characteristics, this principle may be applicable to the structure of various kinds of organs which consist of similar cells. However, as these partial structures are integrated into one individual body, the function of these too must be integrated in the function of that individual. Thus, the functions might be interpreted as various kinds of physiological or psychological phenomena, but from a holistic point of view, they are the phenomena that an individual living thing always expresses, and from the point of view of function, they can be interpreted as the expression for living things to maintain their own structure. However, as structure and function are inseparable, neither structure nor function are conceivable without the other. It may seem that function exists to maintain structure, or structure exists to maintain function, but neither is actually the case. All the structural and functional phenomena are expressions for living things to maintain themselves. In other words, as living things retain their existence, they always express themselves structurally and functionally.

If we think along these lines, we can say that only those having structure and function for life are living things. Living things are those which are alive, living things with life. Further, in the sense that I mentioned before, even the cells that constitute our body are existences with such structure and function, and also are living entities. Therefore, a cell is a living body. Even living things can be regarded as a kind of living body. Nonetheless, why do we not regard a cell as a kind of organism? If there is a difference between a living thing and a living body it should be made clear at this stage of the argument. We must again consider the point that living things originally grew and developed from one cell. If a living thing is an integrated body of multiple cells resulting from the growth and development of one cell, this means that a living thing does not precede its constituent cells, nor do the cells precede the living thing. Rather, a living thing is the aggregation of these cells and that aggregation is the result of the development of the living thing. As such, growth and development means the growth and development of a living thing and it must be the growth and development of both structure and function. Thus, the relationship between a living thing and these cells is the relationship between the wholeness and parts of one integrity which undergoes growth and development in structure and function, and "living thing" is the name given to this one organic integrated entity. Therefore, although each cell is of course alive, when we say one living thing is alive, we refer to the organic integrated exertion of this organic entity. When we say the living thing dies, I think it is the collapse of this integrated exertion of forces. Thus, the maintenance of the organic integrated process which a living thing expresses is the maintenance of life.

Therefore, living things that have biological structure and function always have an organic integrated body and life as the manifestation of their organic integration. Then, the life of a living thing cannot exist apart from its body, nor is there any body apart from life. Body is nothing but for life and vice versa; that is the meaning of live living things in reality. Since even a cell is alive, in this sense it is not wrong to conceive of life for a cell. And since a living thing is also a kind of organism, it is not illogical to combine the life of a living thing and the life of a cell in one notion of organic life. However, the life of a living thing cannot be conceived without its body, in which is reflected the fundamental principle of the inseparability of function and structure for the existence of the organic body.

My view of life is that the life of living things is an attribute of living things. Living things, therefore, are alive and it is impossible to conceive of living things other than as living things with life. However, as the Chinese characters for *sei-butsu* or *sei-tai*[58] indicate, living things are first of all conceived to exist as things. Thus, they tend to be regarded as a physical and morphological existence, so that life tends to be left out, and has to be tacked on with difficulty afterwards. This is because this world is first of all understood as a world of things, and this is in fact known as a natural feature of our recognition. There is, therefore, no point in making an issue of it. We do not know when the word *seibutsu* (living thing) was created, but it is unlikely to be a translated construction because there is no foreign word that is appropriate. Although there is the word "creature," it cannot be conceived without associating it with God as the creator. The word "organism" is a more scientific term; however, although "organism" means living thing, it has, as mentioned, a broader meaning such as living body or organic body. In addition, we also distinguish organic chemistry from inorganic chemistry. In my view, if words like *dōbutsu* (animals) or *shokubutsu*[59] (plants) already existed and were not translated foreign terms, then the word "living thing" must have included both of them in its meaning. Even if there were no equivalent word in other languages, this word "living thing" would have come into use of necessity. So if the word "living thing" carries a meaning which includes animals and plants, it will be clear that although it can be applied to an amoeba, which is a self-contained entity, it cannot be applied to the cells which constitute the body of a living thing.

A question remains, however. A "living thing" originally is an integrated whole, yet the cells that are part of this whole and part of living things are not themselves living things. However, even multicellular living things are the result of the growth and development from one cell. If we go back far enough, living things can be traced to the life of one cell. As such, even the life of living things can be reduced to the life of one cell. If we use the example of embryology, what grows from one cell becomes the body of a living thing and the life of one cell develops into the life of one living thing. In this process of development, to what extent is its life that of the cell, and from whence comes the life of the living thing as a whole? There is no distinct boundary. The organic integrated body which grows and develops with an indistinct cell–body boundary is a living thing. Although a fertilized ovum is only one cell, this is the living thing in the nascent

---

58 *Sei* 生 (living) + *butsu* 物 (thing); *sei* 生 (living) + *tai* 体 (body).

59 *dōbutsu* 動物 (literally, "motion" + "thing"); (*shokubutsu* 植物 (literally, "to plant" + "thing").

stages of the embryo. When people talk about the birth of living things, when the seed sprouts a bud, or the egg hatches, or a baby is born, they regard life as having been added to the body of a living thing. This is just a convenient manner of speaking. In reality, just as the living body does not suddenly appear, the life of a living thing does not abruptly emerge.

It is probably wise to end this discussion on life here. As a concluding statement of this section I would like to ask why the inseparability of structure and function is the fundamental principle for living things or, by corollary, why livings things exist with body and life as an inseparable set. However, to attempt to answer this question is too difficult. I would like to keep it as a tentative discussion. The world where we live is the world where space and time are integrated. In reality, we only know the world in this way. The fact that living things have a form and a structure is an expression of the spatial existence of living things in this world, where everything has a spatial existence. Imagine a world without time. That world would probably be a world with structure alone, with no aging or movement. We do not live in such a world. Can we conceive of a world with time alone and without space? We can conceive of a world without time, but I at least cannot conceive of a structureless world with time alone. That is probably because we are accustomed to perceive this world as consisting of things which have various forms. Of course, a world with only structure is not the world we actually inhabit. The only world we know is one in which everything exists and perpetually changes. This is a world with space and time as one inseparable set. The instant a living thing dies, its body begins to decay. In order for a living thing to continue to live in this world, it must maintain its body by resisting the decaying process. To do so, it must continually renew its body. In this sense, living things create themselves and grow and thus they also live. Although the growth of an individual living thing may cease, within the body old cells are constantly replaced by new ones. Last year's or yesterday's body is not the same as this year's or today's body. This is called metabolism; this created thing continually makes something new. In this way living things maintain themselves in this world. This creating and newly creating phenomenon is illustrative of the very character of living things, which are structural and functional entities.

We can think along the same lines concerning life. A world with space and no time is a world which does not age or move. This may be a world of death, but it cannot be a world of life. When we talk about the world of things, we tend to think of such an abstract world with form or structure only. We know that living things with only structure or body do not exist in our world and they are always structural and functional and have a body and life as one set. This is so because our world is a spatial and temporal world. To state it differently, living things have structure and function as one set and body and life as one set; as such, living things can be constituents of this world, where space and time make up another set. If we consider that structure or body precedes function or life, that is the same as thinking that space precedes time. That way of thinking perceives life and body separately and is like thinking of space and time separately. As living things were born in this world, and grew and developed together with this world, the only mode of existence for living things must be that they are simultaneously structural and functional, reflecting the constituting principle of this world where space and time are an inseparable set.

However, this world is made up not only of living things, but also of nonliving things. Is the basic requirement that everything have structure and function to reflect the constituting principle of this spatial and temporal world confined therefore to living things? Even nonliving things cannot be exempt from the fact that everything changes in this world. When living things die, and can be regarded as nonliving things, how do we interpret the body when it begins to disintegrate? This indeed means the destruction of its structure and disappearance of its function as a living thing. However, the fact that living things cease to be living things does not mean the disappearance of structure or function. The process of disintegration means that the structure of a living thing changes into the structure of a nonliving thing, and so too for function. When it is a living thing, it retains the structure and function of a living thing; but once it is changed into a nonliving thing, the disintegration continues to alter it until it becomes stable as a nonliving existence in its structure and function. Since in the growth of living things the bodily structure is always changing and being renewed, growth is structural and functional. Also, as for disintegration, the body itself is always changing so it is again a structural and functional phenomenon.

Generally speaking, we think that nonliving things do not move or change. We think this because from our point of view living things are always moving and changing, and on this we base our view of the difference between living and nonliving things. However, even nonliving things are not mere entities with structure only, as recent developments in physics clearly demonstrate. First, the structure of atoms, which are material units, is not at all static. Our solar system itself is not static. The sun rotates on its own axis. We conventionally use the term "action" instead of "function" for these movements as distinct from the case of living things. In any case, from the solar system to the atom, wherever we find structure, it never occurs alone, but is always accompanied by action. What does this mean? It means precisely that because space and time are inseparable attributes of this world, it is impossible for anything to exist with structure alone. Therefore, as the inseparability of space and time is the constituting principle of this world, all existence must be structural and functional, and this is not necessarily confined to living things but must also be applied to nonliving things. The inseparability of this structure and function is not only the principle of existence of living things or organisms, but is the principle of all things in this world. This world can be said to be a world of space and time in an inseparable set as well as of structure and function as an inseparable set.

A living existence is something with structure and function, and body and life. However, structure and function cannot be confined to living things' existence. Living things can be characterized as such because they only have body and life. Whether it is called body or life of living things, beings cannot emerge from nothing. Already, as a cell, it is an existence which has structure and function as well as body and life. Even a cell exists by the division of a cell or the fusion of two cells, so that if you follow the process from what is made to what is making it, the origin of living things may be dated back indefinitely. Still they are living things. Essentially, living things can be made only from living things. No one has thought it strange that when a living thing dies, it always becomes a nonliving thing, and that, on the other hand, nonliving things cannot become living things. However, if living things did not exist from the beginning of this world, and if it was originally a world of nonliving things, there are two alternatives for the origin of living things. One is that living things appeared suddenly in the world of nonliving things and so life too came into being then by chance. However, it was only once in the history of the world that nothing was transformed into being. The other alternative is that no existence can emerge from nothing. When we say "nonliving things" and "living things," it sounds equivalent to "nothing" and "being." However, nonliving things are constituents of this world which exist with structure and function. The structure of nonliving things changed into the structure of living things, and the function of nonliving things changed into the function of living things. That is the evolution from nonliving to living things. In this interpretation, even life did not come into being from nothing. The life of nonliving things evolved into the life of living things.

If, however, you admit life in nonliving things, many think that they cease to be nonliving things, or that this is a sort of pantheism. Still, I do not find it a problem to admit the life of nonliving things. Without tracing the evolutionary history here, it is certain that the life of all living things we see now originated from the life of one cell. That being the case, we must admit that this development was analogous to the growth of the body from a cell to a living thing. The growth of the body occurs when we absorb things from the environment. What we absorb is assuredly nonliving things or matter. Our body is being made by assimilating these things. In this case we do not create a being from nothing, but we transform one existence into another existence. As the growth of the body cannot be conceived without the growth of life, and life cannot be understood as beings coming from nothing, and as the growth of life corresponds to the growth of the body, living things take the life of nonliving things. By assimilating this nonliving life, they develop their life. This is the only logical way to think about this.

If you are concerned with differences, then mankind, animals, plants and nonliving things are all different. However, if you look at the similarities, then these are all part of this world and exist by the same basic principle of existence. There is then no reason to confine "life" only to living things, but we can say that there is nothing without life and wherever things exist there is always life. We can then interpret this world as one of space and time, of structure and function, and also of matter and life. Although a discussion of matter or life is necessarily very abstract, in reality there are always differences between matter, such as nonliving things, and plants, animals and mankind. They are also not equivalent in life, but are different kinds of life; the life of nonliving things, the life of plants, the life of animals and the life of human beings. We should not forget these differences. The things that constitute this world are similar in essence, but at the same time they are different. Although they differ, simultaneously we can perceive a prevailing commonality; in this we can glimpse the character of this world, which originally grew and developed from one thing.

3   On environment

The contents of the previous chapter are sometimes confusing and I myself think they were not well written, but they more or less express my thought. Namely, that as we live in a spatial and temporal world, its various constituents have a structural and functional mode of existence, or, in other words, a bodily and lively manner of existence. If this world consists of living and nonliving things, then, in principle, this must be true for both. However, in order for living things to have structure and function and body and life, or even to maintain their existing state, they must constantly renew themselves. Living things maintain life through the continuous creation of new cells by old cells. For this biological process to proceed efficiently, living things must constantly obtain and absorb materials from outside and expel what they no longer need. Although they ingest matter and nonliving things, of this the body of living things is constructed. In other words, through the nourishment provided by the life of nonliving things, the life of living things grows and is maintained. However, I should add the following clarification. An atom of oxygen or a molecule of water, whether inside or outside the body, remains the same atom of oxygen or molecule of water. Since we have already succeeded in artificially producing some of the complex organic compounds found in the body of living things, the time certainly will come when the kinds and combinations of substances that make up living things will be known. Even so, at present to what extent is it possible to make even a very simple living thing artificially in a test tube?

I have not the slightest doubt that there is a materialistic basis for the existence of living things. One plausible view is that this world consists of things, not necessarily living things, and that it is a world of matter. However, the matter of living things is not a mere aggregate of compounds, but is a very complex organic integrated system. The action of the organic integrated body is the expression of life. Thus, the life of living things from this point of view is merely the organic integration of the life of matter, but it does not mean that there are changes in the life of matter. That is, the atom of oxygen keeps its life as the atom of oxygen and life in this sense is nothing but the action expressed by the atom of oxygen. The expression of the integrated organic body is the expression of life and that process controls the movement of the atom. We do not need to invoke the life of matter; the atom itself does not change, but its action is under the control of the living body. Herein lies the difference between the mere existence of matter and the life of matter. The living things as integrated bodies and the life of living things as an integrating process directs the action expressed by the atom of oxygen.

We should not forget that the integrated body possesses a wholeness; if you analyze it, it can be reduced to cells and atoms and electrons but the phenomena that these atoms or electrons express themselves cannot simulate what the integrated body expresses. For instance, it is nonsense to explain why birds fly and fish swim in terms of cells which cannot fly or swim. In this way we can understand what living things or the life of living things means and that there are differences between the two worlds. One is the world of matter or cells which constitute living things or the life of living things, but they are on a lower level and in a different world from the other world of living things which is an integration of them. Physics and chemistry, which deal with matter, developed earlier and independently of biology. At present, cellular research is still included in biology, but in the future we may imagine that cytology will develop into a distinct interdisciplinary field that deals with an area somewhere between living things and matter.

Living things remain living things; they are neither mere cells nor mere matter. An organism is an organically integrated whole of these cells and matter. This seems clear, but we need to ask what in fact is integrating things into wholeness or what governs them? As we always think in human terms, we may think that the agent of integration or control is the brain or mind. But this does not make sense when we consider plants as living things. Some may insist that a plant is a kind of automaton or self-reproducing machine. In such a discussion this choice is a matter of opinion and nothing is certain; we cannot say this *must* be the case. However, in an integrated body or wholeness, there must be one element without which we cannot make things into an integrated body and without which everything disassembles. We may call this element integrity or wholeness; whatever it is, we must suppose there is something. In the presence of this characteristic we will find wholeness, or, if we find wholeness, we will always find this characteristic. Let us consider this integrity more carefully. There must be something in it that prevents parts from doing all that they would like to do, and so in this element there must be a certain principle by which it controls parts. By assuming the presence of this sort of principle it makes sense that the parts are controlled and regulated. To take it a step further, it is really ideal if each part comprehends that principle and accepts its control; if that is the case, there are no connotations of conflict included in words such as control and being controlled.

We can see a perfect example of such wholeness in living things. The cells are constituent units of living things and cannot exist apart from living things. The fact that no living thing exists without cells means that we cannot think of the existence of parts without imagining the whole and we cannot conceive of the whole without assuming the existence of parts. Moreover, living things are not something in which the parts preceded the whole and were combined later, nor in which the whole preceded the parts which later separated. Although everything originally grew and developed from one thing and what was created in turn became creator, the differentiation of cells was in fact specialization in order to perfect the wholeness. So that if we admit integrity in the wholeness of living things and if we can imagine a kind of principle in this integrity, then unless we suppose that this integrity or principle exists right from the first embryonic cells, we cannot understand the genesis and development of living things. Even at conception, before the differentiation of the cell, which has not appeared by chance but is transferred from the parents, the integrity or principle in living things must also be hereditary. In truth, this integrity or principle must have been inherited since the genesis of living things. It has been passed on through propagation and probably living things will continue to transmit this inheritance.

The phrase "organic integrated body" is a descriptor added to the word "organism." Living things are organic integrated bodies, but not all organic integrated bodies are living things. A living cell is also a kind of organic integrated body. In that sense a cell may have its own life course as a cell, but cells are a part of a living thing as a whole and controlled by the principle of the living thing as a whole. Thus we must consider initially the principle of the organic integrated body of living things. What, then, is this principle? The answer is already prepared in what I have discussed, so I would like to draw on this gradually as required. That is, the regulating principle required for the existence of living things cannot contradict the original principle for the existence of living things in this world. Or, it may be clearer to say that the integrating principle became necessary in order not to contradict the fundamental principle for the existence of living things.

In this spatial and temporal world, everything tends to maintain itself in its present state. That must be inherent in the spatial characteristic that is an attribute of this spatial world. However, another, temporal, character of this world tries to resist the maintenance of existing conditions and make everything change. Thus, in the actual world what lives must perish. Even so, in this world living things do not cease to exist nor does the world become a chaotic, nebulous place. That is because this world has structure and function. There is function that maintains the structure and structure to maintain the function. Living things are constituents of this world. Therefore, they are structural and functional and as this world remains spatial and temporal, living things also remain and maintain their structural and functional existence. In order to do so, among living things what is created in turn becomes creator and adds to this world what is similar to itself. Although we may say that cells re-create themselves and thus maintain individuals and that individuals re-create themselves and thus maintain the species, we may suppose that from the point of view of cells, this process is the maintenance of the cells themselves, and from the point of view of individuals, it is the maintenance of the individuals themselves. In short, in this way cells reproduce cells and individuals reproduce individuals and as such they can be reconstituting elements of this world that create and maintain the world as it is. Living things maintain themselves and what is created in turn becomes a creator; if this process is called living then living itself is the leading principle in this organic integrated body. In other words, for living things, to live itself must be the ultimate objective. If I say that to live is the objective, perhaps we will not be completely satisfied, but for the moment let us leave it at that. We may call it maintenance of life, but anyway living things can be living things in this world by living, and this is much more fundamental than the necessity of brain or mind.

Now, if we admit that to live itself is the objective, and that the organic integrated body which controls parts according to the principle of living is equivalent to living things, then in plants or lower animals in which we cannot recognize any brain or mind, how can we find this principle or control, or if we must find some kind of integrity, what is the concrete expression of this integrity? We can see one concrete expression of integrity in that a living thing has its own form. Integrative control or rule suggests a kind of territorial element. There is always a limit or range in control; if there is no boundary, there can be no control. That living things have their own forms and that the living body is an independent system in itself is an expression of the integrity of living things or of the spatial character which attends the integrity. In this sense, even a cell is a self-contained system limited in space by membranes. Because a cell is such a self-contained system, living things likewise may be a self-contained system. Apart from cells, a living thing in general presents itself as an individual expressing one self-contained independent system; this has an extremely important meaning and it seems to me that this independence and integrity cannot be considered separately anymore. Therefore, when we are born and for the first time assert ourselves as an independent system, even at this stage we can see some significance.

As a result of the fact that this integrated being called a living thing is an independent system, we can turn to consider the environment which accommodates the living things. Although an organism is an independent system, in order to live it must first take in food from the environment and find mates there. Thus, it is clear that living things cannot live apart from the environment. In this sense living things are not self-contained independent systems that can exist on their own, but if we think of one system that includes the environment, now for the first time living things can be understood in a concrete form of existence. Living things that are considered apart from the environment are not living things in their reality. Here again I would like to stress that the outside world or environment does not precede the genesis of living things. Even these environments are part of this world and have grown and developed from one thing together with living things. In this sense, living things and the environment are originally of the same kind. This means that the ship does not precede the passengers but they all originate from the same source. In the same way, living things do not precede the environment nor does the environment precede living things. Our world is such that we cannot conceive of the existence of living things without the environment, nor can we conceive of environment alone without presupposing the existence of living things. This must be our world.

What, then, does environment mean in concrete terms? To us it means the outer world which accommodates us. This can be understood in narrow or broad terms for we can include, for instance, the moon and stars in such an outer world. If we interpret the environment in such a broad sense, the world that we can recognize is our environment and the environment is nothing but the world where we live. However, it is unlikely that the moon and stars are understood in the same way by plants and amoebae; plants and amoebae are also unconscious of any environment as their own. So, instead, if we define environment as the place from which a living thing obtains food and also the place where potential enemies exist, this can be understood simply as that living things take food and flee on seeing an enemy because they cannot live unless they do so. That they take food and escape from predators means that living things cannot cease their activity even for a minute. Yet what living things seek is in fact perhaps not activity, but a peaceful life while maintaining constant equilibrium. However, they seek food because without it this equilibrium cannot be obtained, and so this activity can be understood as the effort to obtain that balance. Yet when they take in food, does it really mean that the living things have obtained equilibrium? Unless the food can be digested, they have not obtained it. When the food is ingested and absorbed, does it then mean they achieved equilibrium? Once food is taken in, they must again find new nourishment or this equilibrium cannot be felt. Like the pendulum of a clock, living things attain equilibrium for a moment, but a constant equilibrium is not possible for them. Additionally, that which is reproduced becomes reproducer, and unless they do so they cannot continue to be living things in this world. That living things live means that living things work. The meaning of being alive is simply that living things are working as they always have in the past and will continue to in the future in order to live. In short, living things make a living and they must secure their subsistence.

To say that living things have life inevitably leads to the question, what is life? If we think of a living thing as an organic integrated body, with an integrating principle, we can construe a living thing as an existence whose objective is to live. But if we move from an objective viewpoint to the standpoint of a living thing itself, it is quite unlikely that it would ask what is life or be conscious of the objective of living. To a living thing it would probably be important that everyday life is peaceful. A living thing must take in food from outside, but if it cannot digest something, then life cannot be untroubled; likewise, if it cannot distinguish between its fellows and its enemies, then it is impossible to expect a peaceful life. If that is the case, food and enemies must be recognized as such by a living thing while they are still in the environment and before they have contact with them.

The environment for a living thing is such that the organism's recognition of the environment is to recognize what is necessary for life. It is completely nonsensical for a living thing to recognize the moon or stars without knowing its own food or predators. If we consider this recognition a little further, food is not necessarily absorbed immediately after it is ingested. Thus, from one point of view, the digestive tract might be considered as a part through which the outside world penetrates our body and as such is an extension of the environment entering the body. I mentioned that a living thing is a self-contained system, but in this sense the body in fact handily takes its environment into itself. However, although ingested food is not becoming a living thing itself, it is already in the process of being assimilated and is now treated as part of the living thing and is regulated by its natural integrating principle. Thus, we can consider it as an extension of the living thing rather than as an extension of the environment. And, in fact, this is the way we actually perceive things.

Therefore, even if not ingested, when a living thing discovers something edible in the environment, at that stage already a process begins of assimilating the food as part of the living thing. If we say environment, it sounds vague, but the environment in the process of becoming part of a living thing is in a sense an environment assimilated by the living thing, so we may call it an extension of a living thing. I mentioned in Chapter 1 that when we recognize things, it is in fact an expression of ourselves with regard to that thing and our reaction to it. To a living thing it is first of all important to recognize what it needs. In fact, it must recognize it. Unessential things may not be recognized. What is the meaning of not being recognized? That is to say, unless it is recognized it does not exist. So that recognition is not merely acknowledging something but is in a sense the attempt to make it its own thing and to feel it as an extension of itself. If we interpret it in this way, then the outer world or environment takes on a slightly different meaning. To something making a living, the distinction between subject versus object or self versus the outside world is not so important as we expect. To a living thing the parts of the environment necessary for making a living are ordinarily recognized and assimilated, and the rest of the outer world does not exist. The world of a living thing exists within the boundaries of what is recognized and assimilated, and the organism is the ruling center of that world. To say outer world or environment sounds like something distant but, in effect, the environment is nothing but the world of a living thing; there it finds its subsistence, and it could as well be called its field of living.

When we talk about the field of living we of course can imagine a kind of spatial extension, but the field of living does not mean merely a space for living but is a continuation, a living extension, of the living thing itself. Thus far we have tended to think of a living thing as distinct from its environment or as something in a specimen box. Thus, we think that the field of living can be separated easily from a living thing and think of it as a kind of stage on which an organism makes its living. However, the real living thing is an integration of the living thing and the environment on which it depends and that is the organized system of the living thing itself. It may be relevant to translate "life field" as the field of life rather than the field of living. The word "life" in English is one word, but in Japanese *seimei* (life) and *seikatsu* (living)[60] have quite different meanings. Here I would like to expand a little on my statement that a living thing is an inseparable set of body and life.

---

60 *Seimei* 生命 (life); *seikatsu* 生活 (living; subsistence).

One way of thinking is to restrict a living thing to a system where it is separated from its own environment. Another approach is to conceive of a system with the environment and living thing as one set. Body is, in a sense, matter, so if we consider the environment and living thing from the point of view of matter, there is a kind of continuity between them as mentioned in the example of food. Therefore, if we are courageous enough to regard food candidly as an extension of our body then it is not at all contradictory to think that there is an extension of life in the food. In fact, there are also parts of our body such as hair or nails which we feel are just an extension of our body, or kinds of implements such as dentures or eyeglasses which we come to feel as being part of our body, as if our life system is incorporating them. Implements are more naturally regarded as an extension of our body than is food, because they are made originally to supplement our bodies. I do not mean that our body and these utensils are the same thing. However, when we are talking about body or life, these cannot exist as a self-contained system separate from this world; thus, there is no reason to confine these to an individual body. Of course, body and life are the center of ourselves and they expand into the environment from the center. As such this is a kind of field. In concrete terms, body and life do not have their own clear boundaries. In particular, regarding views of life, I am not content with the conventional abstract theory of life and I hold a view of life based upon a materialistic view, that life cannot be separated from the body. I regard life as extending to this world and because of that extension this world can be our world. That is the best conclusion I can reach at this time.

The environment that a living thing can recognize is the only environment as far as it is concerned, and constitutes the world to that living thing. The organism is the ruler of that particular world. Therefore, a living thing does not integrate and control only its own individual parts, but integrates beyond the individual body to include its environment. This means, therefore, that although this world is one, since various kinds of living things dwell in it, the world is different according to each living thing. Because the world differs, the environment where each lives also varies, and this means that their way of recognizing their environment varies. How, then, do they recognize their environment? How do the lower animals and plants, which do not have nervous systems or sense organs like ours, recognize the environment?

This is a very important point. If I cannot explain it clearly, the points I have made so far will collapse just as I am nearly at my goal. Although I do not intend to write things that I do not know, this important point has not been made clear by biologists. When we examine various kinds of living things, it is evident that there are several grades of development of nervous systems and sense organs which culminate in mankind. However, the fact that these stages exist may imply that in the process of evolution living things expand their own environment, or extend the world to which they can react and where they live. The extension of the environment means, in short, the expansion of the world that they recognize, which in turn means the enrichment and intensification of their integrity. Thus, the phenomenon of mind or consciousness is required only for us who live in such a world but is meaningless to lower animals and plants. However, it does not mean that these animals and plants do not recognize their own world in their own way. Even these plants and animals recognize their respective worlds. Unless this is conceded, I suspect that a true evolutionary interpretation is not possible.

I think that even lower species of animals which live in a comparatively narrow world must at least distinguish between edible and inedible things. Is this distinction made consciously by the animals? In Chapter 1, where I discussed recognition, I stated that the essence of recognition is to know affinity. Then food, in terms of affinitive relationship, is rather foreign, yet since these animals can recognize what is edible and are rarely mistaken, we need an explanation. Food does not precede living things and food and living things originally had an inseparable existence. As such, food is in a very close existence to living things. This is because food is in fact an extension of the body of living things and the source by which they are sustained. Therefore, the relationship between food and living things is not one of biology or taxonomy, but of direct affinity of body to the living things. Because food is an extension of their own body, living things recognize their own food; this means they in fact recognize themselves. Although some of the materials are food, whether or not living things always recognize them as food is open to question. Probably in lower animals they recognize things as edible when they become hungry, but when they do not need them, they are not perceived as food, and they recognize and seek other things as necessary. If you put food and a mate in front of a particular animal, it probably does not feel difficulty in choosing between them. When it needs a mate, food is perhaps not recognized as food. Although it sounds a little strange, even humans seem to respond in a similar manner, though it is not so obvious as in other living things.

Thus, recognition varies from one occasion to another according to the requirement of the living thing. If this requirement means a particular behavior according to the integrating principles of that living thing, then in essence there might be no distinction whatsoever between licking an injured part or scratching an itch or taking in food or avoiding enemies. When living things perform these actions, they are not consciously doing so in order to live, but probably they simply scratch because they feel itchy and eat because they feel hungry. Thus we can say that they use their instinct or instinctive nature. The word "instinct" implies unconsciousness or unselfconsciousness so that in an extreme sense we should probably deny that they *feel* itchy or *want* to eat. The conventional explanation has been that all the activities of living things can be attributed to instinct, but in resorting always to instinct, people reveal that they cannot interpret living things as living things at all. This is not much different from regarding living things as automatons.

Thus, it is necessary to reflect a little more on the behavior called instinct. If we say it is not self-conscious or is unconscious, there are many things even we do in that state. Even when sleeping, if a flea bites, we will scratch. While awake, we are not conscious of all of the movements of our internal organs. One interpretation may be that we were originally conscious of these movements, but as we came to live in an increasingly complex world, we economized by dispensing with the consciousness that we did not need. However, another possible view is that originally there was no consciousness, and gradually consciousness grew according to necessity, leaving places where consciousness is not required as they were. In this case the phenomena of each cell or the movement of digestive organs of which we are unconscious can be interpreted as vegetative, autonomic phenomena inherent in ourselves. In any case, the fact that every kind of behavior that lower animals and plants show unconsciously is not only in accordance with their own objective but also is in accordance with a guideline so that the various actions do not interfere with one other, seems to show that there is something which cannot be explained by instinct alone. So we may say that it is the expression of an integrating principle that living things have by nature. Then at least integrity provides a better explanation than instinct. But this integrity also stands in the way of our understanding. We need to consider this world of integrity more deeply, in order to reach the essential point of the discussion.

Living things control themselves and their environment by their integrating nature. If the environment or the world are nothing but an extension of themselves, the integrity of living things is nothing but the regulation of themselves. In species like us which have developed a nervous system, there is a center of consciousness which represents the center of our behavior or life, so in our case our integrity immediately signifies our autonomy. However, even living things which have not developed a central nervous system or sense organs similar to ours take in food and avoid predators when necessary and do not show any contradictory behaviors. Can we not say that these organisms' integrating character is in fact the same as humans' autonomy and independence? Of course these organisms' recognition of food or predators is different from ours, which is more conscious while theirs is more rudimentary. As we have no detailed knowledge about this, it has been suggested that they react to chemical stimulae. This is a materialistic explanation, but it is nevertheless one kind of recognition and these organisms are in fact reacting autonomously. If they react autonomously we must say then that they recognize autonomously.

If we assume that even in the primitive behaviors of living things in which recognizing immediately leads to action and action immediately leads to recognizing, if these behaviors are in harmony with the integrativeness of living things, this action or recognizing must somehow be felt by them. We cannot explain the sequence of behavior from hunger to satiety unless we see it in this light. Then may we allow ourselves to imagine a kind of latent consciousness or protoconsciousness that might differ from ours? Just as in us the center of consciousness determines our life or behavior, can we not imagine that in these living things too this center of protoconsciousness determines their lives and behavior? If this suggestion is admitted, we avoid the problems of various explanations that regard lower living things as having no consciousness. Whether they are advanced or not, all living things must act. As such the living things in this world must be autonomous. If recognizing and behaving are the same, then recognizing itself is done by them autonomously. And so this autonomy was inherent in living things from the genesis of living things in this world, and consciousness or mind, which has appeared later, was latent in this autonomy. Here we must take the same care in considering the stage before everything differentiated from our viewpoint after the specializations. All living things have grown and developed. Even an egg cell, as yet undifferentiated, was an autonomous existence; each of us has developed as a being with autonomy. As we already assumed life in cells and life in plants, it is not too strange to admit now that cells and plants have their own mind. But of course we must remember that cells are always cells and plants are always plants and that they are completely different from us. Thus, when we talk about mind or consciousness in them, it is very different from ours.

I have come at last to the heart of my narration, but if I had planned better to reach my objective and capture my audience, I think I could have written more skillfully from the beginning. What I wished to relate is that the life of living things, which is continuous action, is the assimilation of the environment and control of their world; that this is, in fact, the development of the autonomy which is inherent in living things. I have not been satisfied with interpreting the life of living things as instinct, which disregards this autonomy, nor with a slightly better explanation through introducing an integrating principle. In order to introduce autonomy, I referred tentatively to mind, but mind itself is one expression of integrity and integrity is another expression of wholeness and independence. Thus, autonomy or subjectivity are characters of wholeness. Perhaps, if I could rephrase it, I would say that living things have a wholeness and autonomy; as a result, we can recognize something like mind in living things.

4 On society

The relationship between life and the environment cannot be described with any thoroughness without considering many other things; however, by introducing the concept of the environment I think that the independence or autonomy of living things has become generally clearer. The environment is the world where the organism thrives, or its field of living. However, this is not meant in a physical sense, such as a living space. From the point of view of the organism, the environment is an extension of itself, which it controls. Of course, an organism cannot freely create and transform the environment. If we regard the environment as something which is not ultimately controlled by the organism, and, in that sense, which opposes it, we can consider that the environment is partly within our body and that our body, which cannot be freely constructed or transformed, is an extension of the environment. As the environment exists in a living thing and the living thing exists in the environment, they are not separate. They belong to one system that originated from one thing. We can say, in a broad sense, that our world as a whole is that system; but from the point of view of each living thing, which is the center of its own world, the organism and its environment constitute one system.

In the system formed by this interaction, the environment can be thought to represent matter. In that sense, even our body can be considered an extension of the environment. By contrast, the living thing represents life or spirit.[61] Thus, to regard the environment as an extension of the organism is to assimilate the environment into the domain of the living thing. Yet questions relating to organisms and the environment have been considered in terms of physical properties of the environment, not in terms of life. This too is a valid method of biological research. A view of nature from the standpoint of living things rather than the environment, cannot be applied

---

61 *Seishin* 精神 = mind, spirit, soul.

indiscriminately to all living things. For lower animals or plants, with whose lives we have very limited empathetic understanding, it is legitimate to express their lives through the physical characteristics of the environment. However, can we say that the view of a living thing which is, so to speak, translated and defined in terms of the environment, truly representative of the actual living thing? Although living things cannot freely create or transform their environment, neither are they entirely controlled by it. Rather, from their respective points of view, they continuously act on and try to control the environment. If living things were simply swept along by the environment, we would not need to recognize their autonomy and independence – they would be nothing more than automata.

Therefore, we must be cautious in interpreting living things from an environmental standpoint. It provides some insight, but we cannot conclude that the environment determines every behavior of organisms. Nevertheless, it is true that there must be a difference from one organism to another; the degree of influence over the environment, or the degree of autonomy or independence from the environment must differ. However, as long as we recognize living things we cannot accept environmental determinism. This is demanded by the fundamental nature of any living thing, regardless of its evolutionary status.

We can understand scientists' inclination to view things from an environmental perspective and to define living things in those terms. However, scientists tend to emphasize the inorganic environmental factors or events that can be expressed quantitatively. Of course, the environment thus expressed, for example, temperature or humidity read by instruments a meter above the ground, is not the environment that grass or insects actually perceive. We therefore now measure the temperature or humidity of the air which directly touches the plant or insect. This is the microclimate or plant climate and it provides a closer reading of the environment. Although we refer generically to a plant climate, the climates of, for example, tall trees or bushes or moss on the ground, are all different. We can also assume that if there is a plant climate, there also must be an animal climate and this must differ according to the way of life of each animal.

However, to what extent are these kinds of inorganic environmental factors perceived by living things? Everyone realizes that a change in temperature or humidity directly affects a living thing and is manifested as a change in its activity. The change in activity indicates that the living thing has perceived a change in its environment. We can say this because, according to the fundamental relationship between living things and the environment, reacting is recognizing, and recognizing is reacting. Let us consider this point. Although we think that we perspire because it is hot, feeling hot is not necessarily the direct cause of perspiring. To say that when we are hot, we automatically perspire, is not an explanation. We perspire as a physiological reaction to the rise in temperature in our surroundings in order to maintain a constant body temperature. We are not conscious of the temperature or humidity of the outside world and do not consciously control perspiring; it occurs automatically. However, our body can perceive physiologically the temperature or humidity of our surroundings; thus, we perspire and our body temperature is maintained. That is, perspiring is our reaction to the environment and at the same time is our recognition of the environment in which we should perspire. It is not necessary that we consciously perceive this. To those who can only conceive of recognition and consciousness as inseparable, this will sound illogical, but this is how we perspire and we will perish if we do not.

Although the evaporation of moisture from the pores of leaves in plants is something different from sweating, the fact that the plant opens and closes its pores in response to the environment also suggests that plants recognize or perceive the condition of the outside world. This has no connection with the development of the cerebrum or accompanying consciousness. To emphasize this, I suggested that even in humans the control of body temperature is not a function of consciousness. Thus, perspiring can be said to be an autonomic characteristic in us. When food is ingested, digestive fluids are automatically secreted in the stomach. That is also an expression of this autonomic property. Even plants such as the butterwort or the sundew make digestive juices and digest insects they have trapped.

Thus, the existence of what I call our autonomic or vegetative property, strange though it may sound, is comparable to the existence of matter, such as air, which we cannot see, or to the action of the internal organs. In plants, which do not move, organs such as eyes are unnecessary. Since plants do not have eyes, they cannot distinguish, in the way that we do, whether or not the neighboring tree is the same species. But even plants can discriminate between same or different species. Plant pollination is principally done only between members of the same species. Although pollen does not have eyes, pollination and hence reproduction would never be achieved if the pollen did not somehow perceive when it reaches the stigma of a flower of the same species. This must apply even in the case of fertilization in animals. Neither sperms nor ova have eyes. In animals, sperms are attracted by a stimulant such as hormones which emerge from the ovum and the sperm move towards the ovum. We can state this another way: the sperms recognize the existence of the ovum. Yet how disappointing for humans that even today we cannot be conscious of fertilization, the fundamental activity for the continuation of life. However, here I simply want to draw attention to the fact that even such things as sperm, which are an extension of our body and carriers of our life, cannot be controlled by our will. Here also we realize that our inborn autonomic character stands firm.

This autonomic nature ultimately belongs to the domain of physiology because it explains phenomena through material, or, more technically, physiochemical characteristics. As a discipline, physiology considers phenomena only from the standpoint of environment or matter. Thus, environment tends to be thought of only as inorganic environmental factors; it is one expression of reductionism which simplifies phenomena as much as possible, treating animals as automata and plants as objects. As one approach it may be fine, but it cannot fully explain the animate characteristics of animals or the vegetative properties of plants. In the same way, we cannot fully explain an individual life by the properties of a cell or explain a cell solely in terms of atoms or electrons. To return to our examples, even if we say that perspiring is an unconscious physiological phenomenon, if we do not assume the existence of a perspiring individual, we will not understand the meaning of sweating. Even if we explain the activity of sperm in physiochemical terms, the existence of the ovum should be thought of as the prerequisite for its activity. Then before asking whether the sperm or the ovum behaves consciously, the existence of the individual which makes the sperm or the ovum must be assumed. Even in, for instance, sea urchins, the dispersal of sperm or ova into the water would be meaningless without a male and female in proximity for the ova to attract the sperm. This also applies to plants that reproduce sexually, unless they self-pollinate.

People tend to think of reproduction or breeding as the preservation or maintenance of the species. However, from the perspective of the individual, it is ultimately the maintenance of the individual. Although an individual dies, the place it occupied in this world is maintained by creating and leaving another which is similar to itself: thus this world is maintained. Of course, I do not ignore the significance of reproduction for the maintenance of the species. Rather, we should note that although reproduction implies both the maintenance of the species and the maintenance of the individual, there is no conflict between them. This is also true for metabolism, which is thought of as individual maintenance, just as reproduction is thought of as species maintenance. The fact that metabolism maintains the individual means ultimately that nourishment maintains the cells which constitute the individual. In this we see why metabolism is originally a cellular, physiological, unconscious autonomic phenomenon. Of course, reproduction also originates in a physiological, autonomic phenomenon. Plants, after all, reproduce. Prior to distinguishing between reproduction and metabolism on the level of the species and the individual, they must first be compared on the same level. Let us consider individual maintenance. Nourishment can be achieved without the existence of another individual only if there is food. On the other hand, the existence of another individual must be assumed to achieve reproduction: not only assumed, but in fact another individual always exists.

This is related to the fact that although the various things of this world are infinitely different when viewed from the standpoint of differences, from the standpoint of similarity, there is nothing that is completely isolated or dissimilar to anything. Why do similar things exist in this world? Monkeys and amoebae, naturally, are not born to humans. The child of a human, furthermore, is not only an individual human, but resembles his parents. This is called heredity. But why do we recognize heredity? As far as the parents' instinct for individual maintenance is concerned, we may think that the more closely the child resembles the parents, the more the objective is achieved. It follows that this contributes to the maintenance of the present state of this world. When we explain why similar things are created biologically, we can only say it is heredity. But if we consider that for anything we name in this world, there exists something similar, it leads me to think that this is a universal phenomenon that cannot be completely explained by biology. If this is so, then its interpretation is beyond my capability, but it suggests to me that in the fact that similar things do exist may lie something like the principle of the structure of the world.

The fact that this world is constantly changing, yet has a structure, suggests that a kind of equilibrium is maintained among its various constituents. If so, then the structure of the body of living things must be one expression of equilibrium. However, there must be various kinds of balances. If we understand equilibrium as a balance of forces, it is necessary, first, that the forces are of the same kind. Second, I do not think that the emergent balance is static. I prefer to think of equilibrium as mutual influences which are balanced, or as a state of interactions. Thus, structure is a relationship produced by the fact that things which have equivalent forces mutually influence each other. As long as it is a relationship, it would be meaningless if there were only one individual. The fact that similar things exist is linked to structure. In this relationship or structure, the relative positions of these similar things emerges as the next problem.

If we apply this premise to the case of living things, we must consider first what "force" means. At the risk of being ambiguous, I think it is necessary to consider the life energy or the expression of the content of living. Then, in this instance, living things which have the same capacity or content of living are individuals of the same species in biology. Then do individuals of a species show anything we might call structure in their mutual interactions?

Mutual interactions among individuals of a species, for most people, refers to reproduction. Certainly, just as two plants must exist within a certain area to achieve pollination, a male and female animal likewise must live within meeting distance. Thus, the origin of social phenomena in living things has usually been sought in reproduction or sexual relationships. Obvious as that appears, it nonetheless may be the case that one male has the same opportunity to meet another male as to meet a female. This is not to emphasize that the proportion of males and females is ultimately the same in many living things, but that such opportunities suggest a reason that has nothing to do with the relationship between male and female.

Let us consider again the concept of environment. The environment is the place where organisms express their content of living. I said that the environment is an extension of the living thing and at the same time the living thing is autonomous and governs the environment. If two living things are balanced in their life capacities, in terms of environment, they do not intrude upon each other; here we can recognize the independence of the individual autonomous organism when we see the environment as an extension of it. Of course, if we consider humans and ants, for example, a human foot can intrude upon the environment of the ant and may kill it. However, two individuals of the same species with the same living requirements, cannot share, as a rule, the same environment. In the case of a threat to survival, such as competing for food, the difference between male and female may be ignored.

Yet, despite the fact that individuals of a species have the most similar requirements and are therefore fundamentally intolerant of one another, why do they exist within a certain area and not scatter? By doing so, they can reproduce, but this does not explain it entirely. That such similar existences generally remain in proximity and not mutually isolated, reflects the fact that they are not independent creations with no relationship, but that they developed originally from one thing. Their differences ultimately reveal their degree of relationship or affinity. Thus, members of a species are found in proximity due to the kinship between them. However, as this kinship reflects similar living requirements, yet makes those which are fundamentally mutually intolerant coexist, there must be something additional to kinship involved. This is also made apparent by the fact that members of a species have the same content of living.

In terms of the environment, having the same living requirements demands having the same environment. If we think of a case where the same environmental conditions exist contiguously, it is natural to think that organisms with the same requirements, even if unable to share one environment, gather and live separately in that continuous environment. We can regard this as the natural result of the autonomous action of such living things on the environment. Here, then, a territorial relationship emerges as another factor binding members of a species.

To say then, that members of a species are bound by kinship and territorial relations is to say that they have a similar form and function and survive in similar ways in the same locality. Taxonomically, a different morphology is usually considered diagnostic of different species, but these differences reflect the fact that their environments and living patterns are different. We cannot say, of course, that morphology reflects all the details of a living pattern. We find organisms which are indistinguishable in form yet which have distinct differences in habits. Yet the day when the details of the lifeways of all organisms are known will not come soon. Thus, we recognize that an organism's form to some extent reflects its way of life. In other words, although the morphology of a dead specimen has a taxonomic significance, from an ecological standpoint, in which the true and original meaning of the form should be sought in its natural living conditions, we always connect the organism's structure with its lifeway. That way, the form is not merely morphology, but fundamentally reflects the organism's way of life. This we call the life form.[62] In my opinion, it is more appropriate not to subsume the details of living in the form, but to consider the form as part of the details of living and to use the word ecology. However, I prefer not to define the word ecology at present, and will use life form for now. Thus, members of a species can be understood as being linked by kinship and territorial relations and which share the same life form.

Having said that the individuals of a species are not scattered but gathered within a certain range bound by blood and territorial ties, there are nonetheless some related forms which, depending on their lifeways, appear to us to be so distant from one another, we can call them isolated lives. Moreover, the gathering of individuals within a certain range is quite unlike the accumulation of trash at the waterside by the action of the environment, or the aggregation of domestic animals which are environmentally confined. Nevertheless, when we look at the dispersion of members of a species, we

---

62 "Life form" is employed here in one of its ecological senses, as "the result of the interaction of all life processes, both genetic and environmental." (*A Dictionary of Ecology, Evolution, and Systematics*, 1982, R. J. Lincoln *et al.*, Cambridge University Press: 140).

find that in every case there is a certain distribution range, just as the individual organism has a living space it needs. This implies that in species, as in individuals, we can recognize the life space of the species. I do not think anyone disagrees with the perspective that the individual organism which is born and dies in the species is nothing more than a constituent of the species. However, that does not provide an answer to questions such as what is the species, or what is the form of the species for an individual, or what influence does the species have on the individual.

The difficulty in thinking about this is that we usually ultimately focus on ourselves as individuals. Even if we talk about the race or the nation, the brains of individuals are doing the thinking. However, the race or the nation were not created by the brains of humans. We cannot say that either the cell or the individual comes first, nor that either the race or the individual comes first. However, the individual is not the same as a cell nor is it the same as the race. One of the biggest dangers in our thinking of the cell or of the race from the standpoint of the individual is the fact that these existences belong to different dimensions. On the one hand, we juxtapose them and think the cell and the individual belong to one set of dimensions and, on the other hand, that the race and the individual belong to a different set of dimensions. If the differences of dimensions in both cases are dissimilar, it is not possible to argue from the former case to the latter. Has anyone considered whether the existence of the race can be inferred from the cell or the individual? From an individual's perspective, the race can be viewed categorically as a unit. However, it is just like the tissues or organs that exist as intermediary units between the cell and individual. Would not the race then vanish as an intermediary unit below that of the human species?

But this is a digression. Let us return to the original issue to consider further the species. As stated before, plants probably do not recognize whether or not the plant next to them is of the same species. This is not simply a function of whether or not they have eyes, but we find no evidence in their reactions or habits to show that they recognize their neighbor. Among animals, however, even in sightless protozoa, we can recognize a reaction to a conspecific – when two individuals are in close proximity, they separate more rapidly. In this sense, I would have to admit that even primitive protozoans recognize their conspecifics. Animals with developed sense organs, especially those with vision, recognize others with their eyes, and, of course, the olfactory and auditory senses also work in this recognition. In fact, olfaction generally seems to play a more important role than vision.

If one can recognize another's odor or can hear another's voice, it follows that the animal can recognize its own smell or voice. It is said that consciousness of oneself is dependent on the existence of another, but this cannot apply to animals that have a lower degree of consciousness. Could it not be that since an animal recognizes an odor as the same as its own smell, and a vocalization as the same as its own, it regards other members of its species merely as an extension of its own body? If an animal exists right by another with quite a different way of living, some kind of conflict in their activities or behaviors may emerge. However, if the animals are members of the same species with nearly identical ways of life, no discordant behavior arises. This can be understood as the stimulus being transmitted without any resistance or that the individuals always behave in accord. We could say that individuals of the same species have a balance of forces, and in this balance between individuals, we can recognize something which is often called imitation.

Therefore, if we admit that living things tend to preserve the individual and maintain the present state, we can also think that they avoid wasteful friction and abhor conflict, and we see a state of equilibrium without friction or conflict, which may naturally result in members of a species aggregating. Thus, even if we do not assume some natural attraction to one another, the reason that individuals of a species gather is that in their common habits they find the most stable, and thus the most secure, life. There, their world is created. That world is the world of the species and the life there is the life of the species. Structurally, although it is a continuous living place of the individual, where it is born, lives and dies, it is not simply a world of structure. It is a system which is part of the world that is continuously being born and developing, and which is spatial as well as temporal, and structural as well as functional. If we now apply the word society or social life to the world of living things, then this term should first be applied to this world of the species, to the shared life of members of the same species. This shared life does not necessarily imply a conscious and active cooperation; rather, as the result of the interactive influence among individuals of the same species, a kind of continuous equilibrium results. Outside of this equilibrium, an individual's survival is no longer assured. The gathering of members of a species is not simply an aggregation, but a shared life.

People often associate the term society with a group phenomenon, but in a broad sense it is a place of shared living for the members of the society. Biologically speaking, it is where the individual reproduces and sustains itself. The relationship among individuals, whether scattered or dispersed, that we can recognize as a social phenomenon, can be a feature that phenomenologically distinguishes types of society, but it cannot be the criterion by which to determine whether or not it is a society. If organisms only reproduced and did not need to feed themselves, then it would, perhaps, be better for them all to gather. Reproduction also can be thought of as a temporal aspect of the organism's existence. Even sustenance cannot be separated from the temporal aspect in terms of an individual's maintenance. When sustenance is considered, the importance of the spatial concept in living things is first made apparent. Animals and plants separated and various life forms emerged due largely to this need for nutrition. In fact, in any living thing the organs associated with feeding and sustenance occupy a larger part than those for reproduction. If the whole life of any individual is considered, the time spent on sustenance is greater than the time spent on reproduction. Was not the first urge prompting living things to have autonomy over the environment and to expand their living space related to the acquisition of food?

Of course, to consider the existence of living things solely in terms of sustenance is as unidimensional and limiting as thinking of it only in terms of reproduction and must be incomplete. If we recognize that a life is a social life which is consanguineously and territorially related to others, then we must abandon thinking of social phenomena in terms of reproduction and evaluate the importance of sustenance in social life. At the same time, from the point of view of an individual in the society, and that which connotes the existence of society, is the distinction between reproduction, which assumes the existence of another individual, and sustenance, which does not. To clarify this point I shall discuss individual sustenance from the point of view of society. That is to say, an organism, related to others in its society consanguineously and territorially, is able to participate in that system as a constituent of this temporal and spatial world by being both a reproductive and self-sustaining entity.

Given sufficient resources, any number of members can gather from the point of view of the society. However, we find both concentrated and dispersed societies because forms of subsistence differ and, ultimately, the food and methods of acquiring food differ. We are reminded of a parallel relation in the lifestyle and social structure of human societies such as hunting, pastoralism and cultivation. In principle, however, human society and the society of other living things are not dissociated. The difference between them, in essence, is the difference in evolution. Predators, in general, are dispersed, as they would be short of food and have to fight over it if they were concentrated. This can also be seen as one expression of the "economy of nature" or conservatism that organisms possess to avoid unnecessary conflicts. Even a solitary life is not, of course, a completely isolated life; in the area where they make their living, neighboring individuals usually communicate. In predators, we can visualize this area which we call a hunting territory. But even predators will cooperate to kill a large animal; thus, if food is always plentiful, aggregation is possible. This is so for killer whales which always live in groups and attack other whales.

By contrast, many herbivores live in groups. Needless to say, their food is plentiful and conflicts over it do not occur. Yet if the group remains in one place, a shortage of food is felt eventually. Thus, for them nomadic ranging is necessary, and at the same time they need enough land for that way of life. Nomadism is often thought of as wandering to obtain water and plants, but it is not the same as wandering. Although nomadic animals move about, they eventually return to the starting point. There is a more-or-less fixed ranging area for a group and therein lies the difference. They have a defined area so as not to invite any unwanted conflicts by invading other groups' ranging areas. Although there are differences between predation and nomadic ranging, we can think of territoriality in this way. However, we should note again that in predation, the territorial unit is usually limited to an individual or at most a male and female pair or a few individuals, whereas in nomadism the unit is a group. If the determining factor were simply whether the animal is a predator or a herbivore, then all herbivores would have to live in groups, while group-living, as seen in hunting whales, would never occur in predators.

In plants, which produce organic substances from inorganic material, there is generally no reason for species members not to live in a group, with the exception of problems such as interspecific competition over space. Our ancestors' success in cultivation depended on that. However, among animals, even herbivores for whom food is supposedly abundant, many do not have a society in which the group is the unit, but appear to live in a dispersed society in which the unit of society is the individual. There must be a reason for this. We can first consider that a typical group behavior as outlined above requires the extension of a certain ranging area, and also that the society in which the unit is the group is only formed when several of that group and their area exist continuously. Writing this I imagine a group of monkeys living in the tropical rain forest canopy, or ungulates living in the steppe or savanna, or a pod of whales in the sea. These are all large mammals, as are humans. We can draw an analogy between their and our social life to some extent, but the analogy ends abruptly for insects. The world in which these small animals live is complex and irregular beyond our imagination, and it may be that in their case there is no continuous ranging area. Where we see a steppe, the grasshopper may see a forest. Even the term forest is based on our way of interpreting the world. These small animals can adapt to such a complex world that means nothing to large animals because, ultimately, they find meaning in their minute environment. Even if they are not predatory, their environmental conditions make gregarious living difficult. For instance, we could easily understand that for parasites an ambitious way of life such as nomadism is not possible. If it sounds strange to imagine nomadism for parasites, consider small herbivorous insects which live on leaves and sap of trees or burrow into the trunk and eat the wood. Their relationship with plants in terms of food can be regarded as a kind of parasitic relationship. As long as they are parasites, they depend upon parasitism and live by utilizing the surplus of the living force of the host.

Herbivores, however, include not only grasshoppers, but also the large ungulates which likewise feed on grasses and leaves. Predators also depend and live on these animals. In one sense the relationship between host and parasite and that between predator and herbivore is the same, in that they cannot consume all the food with impunity. If we usually do not call the latter relationship parasitic, is the difference determined by the discrepancy in size between those who eat and those who are eaten? Certainly, size provides an easily recognizable criterion. But to take this a step further, the fact that those who eat are larger than those who are eaten means that the former are in a position to govern the latter, to extend their autonomy over them and to incorporate them into their own environment. Thus, for the ungulates, the steppe is none other than their nomadic range. By contrast, where those who eat are smaller than those who are eaten, or where the parasitic relationship holds, the parasite does not control the host, or we can say the host's autonomy blocks the autonomy of the parasite. If the parasite overcomes the host, the parasitic relationship vanishes. However, parasitic organisms probably avoided such futile conflict from the beginning in their process of adaptation to the environment. In thus escaping such useless territorial and dominance conflicts which might have emerged between them and various other organisms, by sacrificing a vigorous extension and development of their manner of living, they succeeded.

Although I do not intend to dwell on the relationship with other living things here, there is one point to add. The relationship in terms of size between those who eat and those who are eaten is not necessarily determined by the absolute size difference between individuals. This is seen, for instance, in the case of killer whales that attack other whales. Ants also provide a good example. Although small, ants are not parasites. Their use of various things as living materials shows the richness of their way of life and the breadth of their world. They make underground nests, feed aphids and cultivate mushrooms, which reveals their degree of autonomy or independence from the environment. Other than human society, there may be no other which rivals that of the ants on these points. There is an aspect, however, which cannot be compared easily between the highly developed insect society and human society and in it we can observe the difference between the worlds in which we live.

Setting aside our human society, compared with the society of nomadic mammals in which the unit is the group, ants can be said to have a society in which the unit is the family. In any event, such a society is not a simple aggregation of individuals which may have been the constituents of a more general dispersed society, but, for whatever reason, ants have mutually reduced the distance among themselves and concentrated in one place. They may better be thought of as members of a society of a different dimension. Although we say that ant society is gregarious, it means only that the constituent unit in which individuals aggregate is either the group or the family. These groups or families, however, are dispersed throughout the society. To avoid confusion, it is better therefore to distinguish among individual-, group- and family-centered societies. In some species members are normally solitary and form families only during the breeding season. In others, members live in groups, but change into family units during the breeding season. Without giving specific examples, I will note here that family originates in reproduction. However, if the family is continuous, we expect offspring to be raised and an increase in the number of family members. The constitution of the family is linked, therefore, with sustenance. In other words, without the guarantee of food, the family will not materialize. Ants are successful in this as they have established a division of labor between those responsible for reproduction and those responsible for sustenance. Their increased longevity is probably also related to this. This is also of interest in terms of care of the larvae. Larvae of the family-type insects are dependent upon adults for nourishment, whereas in cases where the adult lays the eggs where the larvae can then sustain themselves, the parent need not remain in the vicinity. The fact that the majority of insects are of the latter kind should be considered in relation to their having, on the whole, a parasitic lifestyle.

I have not yet considered the lives of offspring and parents separately, but in, for instance, insects in which the immature are caterpillars and the adults are butterflies, from the point of view of the life forms, their worlds must be different. The larvae which, broadly speaking, live parasitically are concentrated at a particular place because the eggs were laid there collectively. We can, of course, appreciate the ecological significance of this, but we cannot imagine that there is a mutual influence between larvae on one tree and those on another. The butterfly's world, however, is much wider and we may imagine some mutual influence even among members of the species which live at a distance. This is the only way to explain their specially developed sense organs. It follows that the society of larval caterpillars, as compared with the society of adult butterflies, is a kind of undeveloped society. Even in the society of butterflies the existence of other individuals is not always the issue. It is very subtle in the sense that it is quite unlikely except by chance for there to be a male and female that come to feed on nectar and that mate. If it is not the time when they are physiologically ready to mate, they may not recognize each other's existence. Therefore, though we refer generally to a society of living things, this has various meanings. For example, plants which do not know whether or not the next tree is the same species, or, in the case of butterflies or mammals, the content of each society is different because the content of living of each organism is different. To state it differently, the meaning of the environment for each living thing is different. For a particular individual, another individual of the same species can be thought of as a kind of environment. The meaning of that other individual's existence differs depending on the state of the particular individual, becoming visible or invisible in its field of living according to its state. For humans too, the field of living is constantly changing and, strictly speaking, the meaning we attribute to a particular object is dependent on the specific time and place. In many animals and other living things, this is simpler and, therefore, more potent and active. That is what was meant by being alternately visible and invisible in an organism's particular environment. This is not haphazard, however, but according to the physiological processes already set out in the life history of an organism. Thus, the kind of environment required at particular periods of its life history in a sense determines the kind of object that emerges in the organism's field of living. If it does not emerge at the right time, it is treated as utterly meaningless or effectively nonexistent. In many instances, especially in insects, this response, or lack of it, is called an instinctive habit.

Therefore, to refer in abstract terms to a society of living things may be misleading. Depending on the content of living of particular organisms, even in the mutual interactions between individuals, or in mutual territorial restrictions, there are many cases that we do not recognize. However, in principle, if individuals of a species gather, there emerges a condition that is possible only then. Where this gathering exists over time in a family or a group structure, they must be members of the same species in order to constitute a species. All species are a gathering of conspecifics. The fact that in plants, or even in parasites, each species has a certain fixed area of distribution means that the species is an entity of communal living in which individuals reproduce and feed. Insofar as this holds, I think that in the concept of species there must be something that fundamentally expresses society. In that sense, sociality is a structural principle of this world where everything was born and developed from one thing. Even in the world of infinite differences similar things exist. We can say there is a structural principle because similar things ultimately oppose one another and opposing things finally have to expand spatially. Sociality can be the fundamental character which reflects this spatial–structural aspect and it can be expected to exist in every constituent of this world. Although the society of living things is ultimately where the individual reproduces and feeds itself, I think that this spatial–structural character of the society indicates a deeper relationship to sustenance.

This has been a long and tedious discussion of society, but there is more to add. In order to help the reader clearly understand what follows, I chose first to describe a species as a real entity in this world, or the world of species as a social phenomenon. I will now consider the species at a concrete level. There are probably now more than one million biological species. Yet we may wonder why these many species are not scattered throughout the world, but each exists in a more-or-less fixed range. If the earth's surface were the same everywhere and if there were only one species, it would probably ultimately distribute itself evenly over the earth. However, the surface is not uniform. The distribution of water and land is not equal. In some places there are mountains, in others there are none. Above all, the sun's radiant heat, the energy source for all life activity, can never be distributed equally over the entire earth's surface. Therefore, the earth is fundamentally unequal and inequality can be said to be a predestined character of our world. Still, the world is able to mitigate this inequality to some extent. During the summer more calories of heat are distributed in the polar regions than in areas close to the equator; wind and sea currents moderate temperatures; water vapor carried by the wind produces rain inland, and the action of wind or water gradually levels the land. However, these phenomena will never eliminate the differences and make the world uniform.

Yet it is this very nonuniformity that allows so many kinds of living things to prosper. The question is, how do organisms incorporate this variability into their lives? Here I will give a very general explanation. Imagine an organism with a life form A; if we restrict the meaning of the field of living to its territorial sense and call it habitat type a, then the number of individuals of A which can exist in a certain area is restricted by the number of habitats a which exist there. However, if A finds alternative habitats a' or a'' and succeeds in living there too, then the number of offspring of A which can live within the same area can increase in proportion to the number of habitats a' or a'' in that area. But accordingly, the life form of the organism expressed as A also will change into A' and A''. This is the formation of a species through adaptation to a change of habitat. Suppose that A was not limited to a certain area but found similar habitats and expanded vigorously outside that area. What happens if the habitat was the same but the climate differed? In that case, along with the change of content of living to adapt to the climate, life form A must also be thought to change. In general, however, a change of habitat more strongly affects the formation of a new species than a change in climate. This is probably because a change of habitat has a more direct effect on the life of an organism than a change of climate, and habitat can thus be thought to have a more important meaning. In that sense, a change of diet, for example, from herbivorous to carnivorous, must be one of the most direct and important changes, and, therefore, we can naturally expect the formation of new species corresponding to changes in ways of living.

Thus, when we reconsider the distribution of a species, acknowledging that various organisms coexist within the same area, if they have different life forms, a different content of living and field of living, it means that in fact their life space is different. In short, even if we say they dwell in the same general habitat, by segregating their niches a territorial treaty between them has already been made. However, if we juxtapose animals and plants, or parasites and hosts, or mammals and insects, they already have come to inhabit very different kinds of worlds. Thus, when we think of habitat segregation or territoriality between different species, we must seek it in the relationship between close species which have not yet diverged very far. The only way to maintain an equilibrium between coexisting close species which ultimately try to satisfy similar requirements in the same area, is by both having the same life form and living requirements. This is, in fact, the formation of a society of a species by becoming individuals of the same species. If, however, these things have any mutually incompatible propensities, those with similar propensities would likely gather together, since avoiding unnecessary friction and seeking better equilibrium more likely expresses the fundamental nature of living things. Therefore, if those with incompatible propensities get together on each side and each forms its own society, they achieve the equilibrium they seek by segregating themselves in the same area. In so doing, these two societies are opposed to each other, but at the same time, they are compatible. Plants and animals, host and parasite, or even mammals and insects, certainly do not now oppose each other; however, if they all originally divided and developed from one thing, they must have passed through a condition of this kind of social opposition. I named this kind of society, whose members are opposed to each other and therefore have to segregate their habitats, the same rank society, or the synusia.[63]

---

63 *Dōishakai* 同位社会 (literally, "same rank society"). Although Imanishi translated the term *dōishakai* as synusia in his English writings, his term had a slightly different meaning from that of the original English technical usage. Synusia in ecology referred to "a community of species with a similar life form and similar ecological requirements; a habitat of characteristic and uniform conditions" (Lincoln *et al. ibid*: 242). Both as a term and a concept, synusia is no longer used. However, Imanishi's term *dōishakai* was meant to stress, in addition, the mutual habitat segregation of species living in the same area. A specific synusia is a *specia* in Imanishi's terms.

On society 49

Those organisms which make up a synusia originally had a close affinity, even if they are now mutually intolerant. Except for this incompatibility, they are closely related and very much alike. Their mutual intolerance is a prerequisite for habitat segregation. As for why they became mutually intolerant, they may, as stated earlier, have dispersed into different environments, and through adapting to the new environments, formed various species. From the point of view of the species, each species becomes intolerant of another and opposes it, but from the point of view of the whole, different species are adaptations of one thing to various places. Thus, if we think that the original surface of the earth was not uniform, with various climatic environments emerging due to varying positions of the sun relative to the earth, or the disposition of land and sea, we can conclude that one species evolved into different species in response to this. Therefore, if one organism did not succeed in occupying a particular place, it could have been taken by an organism from a different lineage. Essentially, it is only those originally compatible cohorts which have become mutually intolerant through the influence of their living spaces of which we can say that, although they oppose each other as species, that opposition is actually equilibrium through place. But if those neighbors of habitat segregation belong to different lineages, their relationship is no longer an equilibrium and the result is all manner of conflicts and disorder. Synusia thus is a structure consisting of the gathering of individual societies of living things and is also a kind of communal society. These societies oppose one another and are mutually intolerant, but each maintains a mutual equilibrium only when the others exist and frame its boundary. Thus each accomplishes the organization of its society, and in this sense they are complementary. So an individual, which consists of an aggregation of cells, a society, which consists of a group of individuals, and a synusia, consisting of the gathering of societies, are each an element of the respective systems that are based on the same principle and that together constitute the world of living things. Although they are elements of systems based on the same principle, the systems are related as parts of a whole. Thus, they must be thought of as elements of systems on different levels.

This must sound very abstract and abstruse. I said earlier that the number of extant species probably exceeds one million. That means there are more than a million kinds of societies of living things. If so, how is this great number of societies arranged, into how many synusiae, and how are they placed phylogenetically? The difference of affinity between plants and animals makes their societies dissimilar, but even in the less developed society of plants, the division into synusiae is evident. This is an interesting fact. Moreover, among plants with little or no mobility and therefore less demand for living space, individuals with similar life forms in general seem to be gathered. The result is that each plant society has a unique character, which may be described as physiognomic. And because the social expression of plants is in terms of characteristic appearance, plant synusiae can easily be discriminated by their general outward appearance.

For example, suppose there is a forest here. It is a forest because of the existence of tall trees. Probably these trees make up a group. The forest does not only consist of trees, and if we enter we see shrubs and undergrowth beneath the trees. Penetrate to the ground surface; often there is a layer of moss or lichen. Imagine another forest also exists here. There may also be layers of trees, shrubs, undergrowth, moss and lichen. In these cases, individuals occupying the tree layers have similar life forms, being in similar living spaces, and can be thought to occupy a similar social status. The same can be said about the other layers of the forest. Therefore, in this world the layer of trees itself forms a synusia and the layer of shrubs, irrespective of the layer of trees above it, forms its own synusia. The tree or shrub layers may be thought of as occupying different stratified statuses. However, when the shrub layer occurs within the forest, the social space formed by the shrubs is seen to be part of the social space formed by the trees. In other words, the society of shrubs is contained within the society of trees. In that sense, the society of trees is dominant to that of the bushes. Since we perceive the group life of plants by their general appearance, we describe them as, for example, forest or grassland. However, from an ecological point of view, the forest is a group of plant societies in which the society of trees occupies the most dominant status, and the grassland is a group of plant societies in which the grass society occupies the most dominant status. The terms tree or herb refer to life forms of plants. Therefore, if the society of trees or of herbs represents a synusia in plants, then that synusia is the same as a life-form society and a life-form community. Plants are classified ecologically into about ten life forms. These vary considerably in size of distribution, but we may for now estimate the number of plant synusiae by the number of life forms.

However, in animals with more complex lives than plants, the classification of life forms may not be easy. Inevitably, we cannot perceive the overall appearance of the life of many small animals leading a kind of parasitic life hidden among the plants. But here, there is a clue for the classification of life forms. Notwithstanding the fact that the Mongolian gazelle and the grasshopper are both animals that live off the grass of the steppe, the fact that the society of the larger gazelle occupies a more dominant status than the society of the grasshopper, and that they do not form a synusia, can be inferred through analogy with the example of plants, in which the tree society occupies a dominant status over the shrub society. If so, would a species of gazelle the size of a grasshopper have to form a synusia with grasshoppers, or would grasshoppers as large as gazelles segregate their habitats in order to live in the same synusia with gazelles? Happily, as we do not find diminutive gazelles or gazelle-size grasshoppers, the formation of an animal synusia is in principle regulated by the affinity between the species. That is, although the life of animals varies from species to species, those with a close affinity have similar lives or a similar life form. Therefore, animal species which form a synusia often are limited to members of a taxonomic unit such as a genus, family or order. The biological debate about whether the species is real or abstract continues even in this century. However, in this kind of example, it is not clear that even the genus or family, let alone the species, are, in terms of synusiae, real taxonomic units. This seems to me to be an area that taxonomy and ecology should develop together.

The synusia can thus be applied in a broad or narrow sense. Among even similar life forms, there is a gradation from the very similar to the less similar. This is based on the affinity existing among things which originally separated and developed from one thing, as discussed in terms of similarity and difference in Chapter 1. As a general rule, the closer the affinity the more similar the life form and the clearer the habitat segregation. Especially when the environment is comparatively uniform, habitat segregation emerges more clearly. For example, compared with the forest in which there is a dominant tree society, the society of plants in a mountain stream is very simple. There, a kind of moss society barely clings to the surface of the pebbles, as with the single lichen society furring the pebbles and stones on the mountaintop. Animals living in such an uncomplex environment, where there is no overlap of plant societies, must likewise segregate their habitats two-dimensionally, or over the surface of the ground. Thus, the habitat segregation which directly corresponds to this kind of environment typically appears as a zonal arrangement. Likewise, we can recognize a zonal habitat segregation in animals living on the seabed. Plants live in a relatively monotonous environment in comparison with animals, and they most often present a clear band-like distribution. Many of the animal species inhabiting mountain streams or sea coasts are adapted to obtain food brought to them by the stream or waves and do not actively seek food to the point where they risk being swept away by the water. Thus, they assume a sedentary, plant-like life which also show clearly the typical banded distribution. Compared with these aquatic animals, the habitat segregation of many terrestrial animals, in accordance with the complex structure of the respective multi-layered societies of plants, is not limited to such planar zones. Thus, some seek a dwelling place in the grasses, others on the surface of the land, and still others in shrubs and trees. However, because they are separated, should we think of grass-dwellers and tree-dwellers as separate societies, as in the case of grass society and tree society which have different levels of society? One school of American ecology classifies animal societies according to the habitat. However, this is based on our human perception of the environment and does not consider the animals' point of view. Their theory is not founded on any evidence that shrub-eating insects or consumers of tree leaves distinguish between shrubs and trees as humans do. Rather, the difference between these insects can be compared to the difference between tape worms in humans and tape worms in dogs, and both cases understood as synusial phenomena occurring through habitat segregation.

Species with dissimilar life forms and more distant affinity living relatively close to each other may at first appear to coexist, but they are not examples of habitat segregation. The variation in their life forms reflect, for instance, differences in methods of food acquisition or other activity. Thus, in accordance with the content of living, the worlds in which they live also differ. Although in our perception their societies may seem to overlap, they are actually separate since the worlds they live in are different. We can ask, then, do these worlds interact? They probably do, and there are negotiations on many points between those living in the same place. It can be thought that those worlds intervene with each other and both share a part of their worlds. Despite this, they can coexist because there is a kind of regulation of the field by which an equilibrium is maintained. For example, whereas between those with very similar life forms and equal forces this equilibrium can only be attained by a kind of spatial opposition through habitat segregation, between those with different life forms and unequal forces, one always occupies a dominant status and the weaker always gives away its position. Here, a different kind of equilibrium can be maintained. Further, although the subordinate gives way in a face-to-face confrontation, that only occurs if one occupies a dominant status. However, the mere fact of their coexistence suggests that on other occasions the subordinate will maintain its autonomous status. Of course, this regulation of the field is due to various factors. Whether or not small ants will catch and kill a large insect, given the condition of the insect prey remains the same, is determined in a particular instance by the number of attacking ants. There are a number of examples in which animals segregate their habitats temporally to gain access to the same food, as in the butterflies and flies that wait for an opening to suck tree sap only when the powerful hornets are not there. This can also become a diurnal and nocturnal habitat segregation. More competitive life forms must avoid coexistence to exist synchronously. These tend to occupy the same place seasonally, and thus employ a seasonal habitat segregation, which is a kind of temporal habitat segregation. Here we see that the synusia is not only related to spatial extension but has an aspect of continuity. The synusia, then, is not simply a structural unit but concretely embodies an organizational constituent of this world as an integrated form of space and time.

Even those with different life forms and which inhabit different worlds share a part of each other's world. They may prey on the same food or live in the same area. Nonetheless, it is a mistake to think these are always members of the same synusia. For example, the tree, shrub and grass societies living in the same place and seeking similar nourishment do not belong to the same synusia. Likewise, we should not think that gazelles and grasshoppers which both live in the steppe and eat grass belong to the same synusia. Therefore, in a synusia that is one step expanded from the essential or primary synusia, the members have both mutually conflicting and complementary lives and do not coexist. In a primary synusia, the insects are connected by affinity with other insects, and mammals are connected by affinity with other mammals. The connection through affinity means that they have similar life forms. Therefore, the primary synusia is ultimately based on life forms that are a community of similar life forms.

Thus, the species of insects living on pebbles in a mountain stream each belong to a society which constitutes a synusia. In that sense, each society belongs to a different synusia. However, we can also say that together they form a life-form community in which they segregate habitats in the same place. This can be considered a synusial existence, at least in comparison with other life-form communities in the same mountain stream, such as fish. This kind of life-form community is also seen as a synusia, but distinguished from the primary, mutually conflicting and mutually exclusive synusia, as being a structure that is based on the primary synusia. If a name is necessary we can call this a synusial complex,[64] but the word synusia gives an ideational, theoretical impression. The term "community" may provide a clearer and more concrete connotation. There should be no problem in calling, for instance, the synusial complex formed by insects an insect community. In terms of social structure, however, it is a synusial complex.

---

64 *Dōifukugōshakai*, 同位複合社会, literally, "same rank composite society." In his publications in English, Imanishi used "overlapping synusiae" for this term.

Continuing along these lines, what we previously recognized in plants as a simple synusia, such as a shrub or tree society, must be, in reality, a shrub or tree community, which is itself a complex of synusiae. It is rare for such a life-form community to consist of only one species of shrub or tree or for it to be identical to the primary synusia. More commonly, various species gather and form a synusial complex. Accordingly, when we analyze these various kinds of shrubs or trees, we expect them to be divided into their own respective primary synusiae. A tree society, which is a life-form community, is a synusial complex that is analogous to the life-form community of insects living on pebbles in a mountain stream. This point is likely to be overlooked because the diversification of plant life forms, in accordance with their lives, is very simple, and moreover their life-form community always appears to be rather uniform to us. If deciduous and coniferous trees coexist, we immediately perceive it as a single society, but the broad-leafed trees, needle-leafed trees or bamboo belong to different phylogenies and have different affinities. If we further analyze them, we will ultimately separate them into their corresponding primary synusiae. Therefore, their development of the same life form and formation of a life-form community is simply a kind of convergence like that seen in whales and fish.

In the case of plants, trees form their own life-form community, but in animals, do whales and fish actually make one? The concept of life form is based on the morphology of living things; from a likeness of form, we infer an analogous content of living. I have no intention to reject that here. However, we have to be very cautious about morphological convergence in animals, because their total content of living is not expressed in their physical form. In plants, in which the evolutionary breadth, so to speak, from a lower grade to higher grade is narrow, the formation of the synusia may to some extent conform to the life form that is based on morphology. However, in animals the breadth of evolution is wide: even within the vertebrates, there is a considerable distance between fish at a lower evolutionary grade and mammals at a higher evolutionary grade. A simple comparison of parenting behavior between fish, which lay eggs and leave them, and whales, which suckle their offspring, reveals that physiological and psychological differences cannot be ascertained by mere similarity in appearance. Moreover, the difference between these species' worlds is ultimately determined by the physiological and psychological differences between them. Therefore, if we try to include morphology in the concept of life form which corresponds to this difference in the content of living, it is not enough to consider only the superficial physical form. We ultimately have to consider the difference in structure, that is, the difference of organization. This, then, is why the formation of synusiae in animals differs from plants, and it is always based on phylogeny or the relationship of affinity. Therefore, despite their morphological similarity, whales and fish do not form a synusial existence together. The world of whales is probably closer to that of land mammals than of fish.

Even so, whales and fish do not live without any relationship or interaction. Probably their similar appearance reflects the fact that their lives do not resemble those of the morphologically dissimilar octopus or shrimp or shellfish. In that sense, can we not think that whales and fish constitute a vertebrate synusia? Or can we not think that land mammals such as elephants and mice form a synusia of mammals? I think these are very pertinent questions. When we focus on similarity, organisms that are related through affinity are ultimately similar, and when we focus on difference, they are different. As long as the concept of synusia is based on affinity, there is no reason to reject this interpretation. However, if we return to the original concept of synusia, the term can be replaced by, for instance, vertebrate community or mammalian community. Moreover, in that sense, the organisms living on the earth's surface together constitute one life community. If we divide these lives, according to the life form, into animals, plants and unicellular lives, we then say that they constitute respectively the animal community, plant community and unicellular life community. Of course, it is unwise to use a word like community loosely. However, given that the present variety of life has divided and developed originally from one thing and has made one another's living possible through habitat segregation over an area or in a single spot, the use of the word community here is justified.

At the risk of being repetitive, if we return to the starting point and try to describe the aforementioned community in terms of society, then it is also a societal complex. I do not regard these communities as synusiae, but call them societal complexes to distinguish them from synusial complexes in which the member societies are on the same level.[65] At this point we can ask again how we determine whether these societies are on the same level. An interim explanation for this kind of social structure has customarily been sought in the domain of food sources. A "food chain" is an ecological term meaning a continuous mutual relationship according to an order of predation. For example, aphids in a rose bush are eaten by the larvae of the ladybug, the spider eats the ladybug, the frog eats the spider, the snake eats the frog, and the eagle eats the snake. Basically, the predator is bigger and

---

65 For instance, various species of trees in a forest constitute a synusial complex in which member societies are on the same level.

stronger than the prey, but in terms of fecundity, as expressed by the number of individuals, the prey species is always more numerous than the predator species. Therefore, the predator can be considered to be in a dominant relationship over the prey, but it can also be thought of as a dependent relationship, or as a parasitic relationship in a broad sense. Any of these interpretations can be justified. We lack precise knowledge of how this relationship between predator and prey in the food chain emerged, but at the present time among existing animal groups this relationship is found among those related by affinity. That is, insects eat other insects, fish eat other fish, and likewise for birds and mammals. This suggests the following hypothesis, at least for now. For the sake of simplicity, suppose that the surface of the earth is entirely uniform. Assume that only one species exists there, and it begins to reproduce at a certain fixed rate, resulting in saturation of the surface of the earth. At that point, the reproductive rate of that species may decrease to a rate which maintains this saturated condition. If change occurred like this, it would be fine, but I feel it is a passive strategy unlikely for an organism. However, if the original reproductive rate were maintained, this life would be reduced to a battle scene of a so-called struggle for existence. This too is unlikely for organisms which avoid unnecessary conflict. Thus, the harmonious solution likely employed by organisms is a division of labor by which, without decreasing the reproductive rate, they maintain maximum numbers by dividing their society into predators and prey. Through this transformation from quantity to quality, so to speak, the absolute number of living things on the surface of the earth has increased. One society became two societies, and by that one species became two species.

For simplicity's sake, I assumed a scenario in which the earth's surface was invariant, but we have also seen that as the surface is in fact variable, habitat segregation became a possibility and synusiae developed. Regardless of the uniformity or otherwise of the earth's surface, this kind of division and formation of species is possible through a food chain. Moreover, if living things actively maintain a reproductive rate as this hypothesis suggests, in order to maintain the societies of living things in equilibrium, the relationship between those who eat and those who are eaten must be an indispensable social function. Although the separation between predators and prey can also be interpreted as a kind of habitat segregation, it differs from the primary synusia of single species in that both species have to coexist from the beginning. Therefore, even though this habitat segregation is the basis for the formation of a synusia, it is in fact the incipient formation of a synusial complex. Here, the link between the primary synusia and the synusial complex is clearly recognizable. I expect this is why the predators and prey have a close relationship of affinity. Because they were originally closely related and in many cases had a synusial relationship, the predator–prey relationship is not a good criterion by which to measure the difference of societal levels.

Thus, we must look for the criterion in simple characteristics such as body size or weight. The fact that modern biology has not yet understood the precise ecological significance of societal levels is somewhat surprising, but my guess is that the clue to its resolution may lie in such simple characteristics. Inevitably, some will doubt that differences in societal levels can be expressed quantitatively by body size or weight. They probably believe there can be no discontinuities in nature. If we gathered all the plants and animals in the world together and arranged them in one place, we might indeed see no discontinuities. However, as I mentioned, more than one million species are never seen everywhere in the world. Only certain species live in a particular area. Thus, biogeographers have attempted many times to categorize the earth's surface into various life zones based on regional habitat differences of species; however, most of them were taxonomists who simply classify species. Probably due to their unfamiliarity with ecology, few, if any, considered the content of the fauna or flora, or why, for example, particular species exist only in a certain area, or by what mechanism these species coexist. These kinds of questions thus fall to the new science of ecology. In these investigations, there must be a fundamental reason why animal ecologists have lagged behind plant ecologists. In fact, the well-ordered appearance of plant society gave plant ecologists an advantage. Anyone doubting this should visit a forest, especially a virgin forest. Even before classification of life forms was widely discussed, the orderly arrangement among tree, shrub, herb, and moss communities made it relatively easy to decide which plants in the forest belong to which societal complex. Those who say, however, that existing trees were not always trees, but went through different stages of growth, such that shrub-size trees and so on up to full size trees must coexist, and that this represents a continuous and healthy virgin forest, do not really know the primary forest. The coexistence of different stages of growing trees is recognized only after the destruction of the forest. The fact that a stable forest shows an orderly arrangement of synusial complexes to the end means that there is discontinuity in nature. If we were to align together all the trees and shrubs of the world, the transition from shrub to tall trees might become continuous, but in actual existence the synusial complexes of trees and shrubs are discontinuous. This discontinuity is directly expressed by the disjunction in size between shrubs and trees.

Animals present the problematic case. I previously noted that even animals inhabiting the same steppe, such as gazelles and grasshoppers, are at different societal levels. This is not only a problem of size, but the fact that there is no grasshopper as large as a gazelle nor a gazelle as small as a grasshopper reflects the difference of affinity between mammals and insects. A societal complex of animals has first of all to be a life form community which is connected through affinity. Thus, there are among the mammals small species which are only about as large as grasshoppers, such as the harvest mouse and the shrew. Yet even those would not likely be regarded as constituting a synusial complex with grasshoppers because of the difference of affinity. In plants, we can lump them together and think of them as a group, but in animals we need first to separate a group by affinity and consider the phylogenetic community. This is where the difference of evolutionary diversification in plants and animals matters. That is, if we take the example of a forest, which is a plant community, among its constituents are moss and lichen living on the surface of the ground and on tree trunks. These belong to considerably different phylogenetic lineages and should be considered separately. However, other constituents of the grass, shrub, and tree societies there have basically similar contents of living and require similar living conditions. In that sense, the fact that even among closely related species such as in the rose family, one finds both tree and shrub species, or in the legume family, we find tree, shrub and herb species, is evidence that these can be regarded as constituting a phylogenetic community. Even within a single species some members become shrubs and others become trees depending on environmental conditions. Thus, although each of these three types of life form appears as a distinct society level in the forest, together they express an evolutionary community which has a certain phylogenetic breadth. Altogether, the evolutionary diversity of plants is comparable to that of mammals, but the mammals constitute only one part of the whole evolutionary community of animals.

From this point of view, within a circumscribed area, regardless of the variety of mammals in a community, we can think there is a clear discontinuity in size. By regarding this discontinuity as a reflection of different societal levels, size can be used as the criterion for distinguishing between synusial complexes. We have to some extent come to think of the structure of the animal community from the same standpoint as the plant community. We can state confidently that elephants and mice are not the same synusial existence. But unlike plants, closely related animals do not spread over different societal levels because their living environment is more complicated. In the process of extending its autonomy over the environment, the autonomy of the animal is affected by the environment and by this process the autonomous organism adapts to the environment and is divided into species. That is why in animals, which unlike plants are spread over different societal levels, the separation of the species is more apparent. It is interesting to note that among mammals, carnivores tend to spread over different societal levels which corresponds to the relatively small degree of environmental determination in their organization. Even so, how shall we explain the discontinuity in size of animals belonging to one evolutionary community and our recognition there of different societal levels? In a forest, the orderly arrangement of members of the synusial complex is probably the most logical; they avoid conflict, and as many as possible have access to adequate light or nourishment. This must be the internal equilibrium of an evolutionary community. If so, does the same apply to animals? Let us consider whether this can be dealt with as an extension of my previous hypothesis.

Through the division between predator and prey in one place a synusial complex was created. However, this means that they arose and developed originally from one thing, and predators cannot survive without eating and prey cannot survive if they are eaten. Predators become larger and stronger to ensure that they can eat. However, prey species also become larger and stronger to lessen their likelihood of being eaten. The result can be thought of as a kind of see-saw in which both predator and prey become larger and larger. Those who cannot become larger are probably most easily eaten. But what about becoming smaller? To some extent a small animal can conceal itself from a large predator. But then the predators may also become smaller in response. Among those who become smaller and perhaps succeed, there must sometime be a division of labor again between predator and prey. This follows from the earlier hypothesis. In this way, one synusial complex is divided into two synusial complexes. At the same time, the number of species increases as well as the absolute number of individuals.

Two synusial complexes formed in this manner are in a kind of equilibrium in their relationship, a balance based on discontinuity. This equilibrium becomes possible through mutual nonintervention. This is a characteristic discontinuity in terms of social structure. For example, if organisms that had become large attempted to become smaller again, or vice versa, even if they remained in an area of mutual avoidance while transforming their size, they would not belong to either synusial complex. An intermediate existence is not permissible as it would conflict with the formation of a synusial complex. Thus, future divergences must be in one of the two directions already given – a new synusial complex is formed when those who have become large increase their size, and those who have become small, further decrease their size.

In reality, the surface of the earth is unequal; but though this simple explanation is not sufficient, I think it contains some truth. The majority of living things on earth are animals. The number of terrestrial animals is four times that of marine animals. This can be related to the inequality of the surface of the earth, according to which living things forming synusiae or synusial complexes have continuously reproduced themselves as efficiently as possible. I discussed discontinuity in nature only in terms of the differentiation of synusial complexes. However, discontinuity is already reflected in phylogeny itself. Thus, if we think of a mammalian community or an insect community, although the latter is a phylogenetic community, it still contains evolutionary diversity. We think all insects are small because we compare them with animals such as mammals. But if we compare small and large insects, the phylogenetic difference may be larger than that between mice and elephants. Even so, is the difference in quantity the same as the difference in the quality? In animal taxonomy, insects and animals occupy the same taxonomic status. But in terms of the ecological content, the difference between ants and grasshoppers is much greater than that between mice and elephants. Previously, I said that the insects as insects form a synusial complex. That probably should be corrected, not from the point of view of mammals, but from the point of view of insects. I think that the discontinuity of affinity is a clue for analysis. Also, the discontinuities in body size, weight or momentum of the members of a phylogenetic community are clues. The mutual relationship among them in the place of their actual living must be investigated and clarified in terms of social structure. As I have become dissatisfied with the contemporary vogue of looking to extreme holism for an answer, and the superficial use of the term animal community, I have attempted as far as possible to systematize it and to regard the world of living things as a concrete expression of this world. However, I am still conscious only of its shortcomings.

5   On history

The various things of this world are not random, unrelated existences, but are all constituents of one large holistic system. In the last chapter I discussed the society of living organisms in some detail in order to illustrate the world as a single structure composed of these elements. To review only the major points, the individual living thing is a constituent of a species society in which it is born, lives and dies, and it is distinct from other individuals of the same species. The species society itself is one constituent of a synusia and is distinct from other species societies. Both the species society and the synusia ultimately have their foundation in a kin relationship. In the structure of this kind of phylogenetic community, a basically temporal development has become spatial; I regard this also as one mode of diversification of living things. In contrast to this, instead of temporal things becoming spatial as expected in the synusial complex, those which should become spatial have become temporal. This was considered to be another mode of development of living things. Therefore, we cannot consider the basis of the synusial complex to be restricted to a phylogenetic relationship, but must recognize that a territorial relationship also already exists there. Where a synusial complex further develops and separates into several synusial complexes, their relationship is based on severance, but it is a breaking of blood ties; the territorial basis of the society's constitution is not lost. Instead, as kin relations become weaker, the territorial foundation would become more clearly recognizable. What we recognize as concrete communities of living things can be seen on closer analysis to be several separate synusial complexes or synusiae, but the whole society is always recognized by us as this kind of territorial community of living things. This is why ecology attempted primarily to classify communities of living things geographically or physiognomically.

This kind of territorial community of all living things is nature as we see it. On the one hand, it is the ultimate society, composed of the individual, species society, synusia and synusial complex; in that sense it is the only whole community of living things. But how should we interpret this kind of whole community? Does this wholeness mean the same thing as that of the species society or synusia? To begin with, the individual living thing is a complex, organic body. The whole cannot stand alone without the parts, nor can the parts exist without the whole, and the life and growth of living things lies in the maintenance of this relationship between the whole and the parts. Because of the inseparability of the whole and its parts in living things, such that each part contains the whole, the wholeness of an individual organism is always expressed as its autonomy. Therefore, the development of wholeness is the development of autonomy. Although a control faculty such as consciousness or mental operations has been one effect of this development, even where consciousness is not recognized, we cannot deny the autonomy of the living things. The wholeness possessed by an individual plant is also always expressed as its autonomy. A holistic thing is autonomous, and an autonomous thing will in some sense or other create itself.

Now, a species society made up of this kind of individual can be considered as a foundation for the individual, but originally, neither the individual nor the species existed first. That being so, the relationship between the individual and the species is also a relationship between the part and the whole, and displays a self-identical structure. Therefore, we may assume the autonomy of the species in its wholeness. The species must also be something which creates itself. Its origin must be in the species itself. Of course, I am not forgetting the environment. I am thinking from the standpoint of the species which has autonomy over the environment. However, the species society of living things is spatial, as the place which the individual, the family or the group occupy is spatial, and in that sense, is morphological. With the exception of these, the species society is rather poor in the expression of wholeness or autonomy as compared with their expression in the individual. This sort of interpretation is surely backed by our intuition. Probably the individuals which comprise the whole species are not as tightly connected as are the parts that make up a whole individual. Though we say that the parts do not exist apart from the whole, and the whole cannot exist without the parts, in the case of the individual, though the part is contained in the whole, in general it is not possible to say that the whole is contained in the part. That is probably because people focus too much on the fact that the individual is reproduced from the reproductive cell which is a very special part. But this is very important for another reason. The individual reproduces from the reproductive cell and ultimately that part is differentiated in the individual. The development of the wholeness or the autonomy of the individual corresponds to this differentiation of the parts. Here we are reminded that where the autonomy of the individual develops, the individual extends its autonomy over the environment and is the center of its world. By contrast to this, within the species a differentiation among the individuals which compose it generally is not recognized. The individual is contained in the species; at the same time the species is contained in any individual. From any individual there is the potential to create the species. The individual is the species and the species is the individual. The species is not necessarily dominant over the individual. But the family or group in themselves can be a species. Animal species in which the family or group is dominant over the individual, have already come a step closer to human society. But in human society a division of labor among individuals emerges. The development of a division of labor, by developing the society, can be thought to have encouraged the development of the wholeness or autonomy of the society. We can recognize here a kind of parallel phenomenon between the development of wholeness in the individual and its development in human society; but in the species society of living things, in general, a differentiation or division of labor among its member individuals cannot be seen. It is a monolayered society which is simply an aggregation of like individuals. By itself it is no more than an undeveloped society which does not have systematic completeness.

The species is represented in the individual and does not necessarily occupy a superior position to the individual. That is simply a given condition of organisms' lives. But it must be cautioned here that in principle and in theory we ultimately have to recognize the dominance of the species over the individual. When organisms that ordinarily live alone aggregate for whatever reasons, naturally, the dominance of the species can be recognized, but even so, it can never be thought that all the individuals of the same species aggregate. Therefore, when we come across an isolated individual or group, this itself appears as a species because of its discontinuity; but when we broaden our perspective, this discontinuity resolves into a continuity and one species occupies a distribution range that supports all of its members. Where a species segregates its habitat from another same-level species, which should be opposed to it, we must recognize the wholeness, and consequently the autonomy, that the species possesses. If we attribute wholeness or autonomy to the species exteriorly in terms of the environment, it can generally be recognized in every living thing. However, there are two kinds of discontinuity shown in the distribution of individuals of a species. One is discontinuity due to an irregular environment, as exemplified by the distribution of small animals which are adapted to minute environments. Their discontinuous distribution is due solely to the discontinuity in the environment. The other is common in large mammals, either carnivores or herbivores, and may be described as a discontinuity within a continuity, in the sense that the member individuals segregate their habitats on their own within the distribution range of the species. Thus, this kind of difference ultimately emerges due to the different autonomy of each individual animal over the environment; the so-called species' autonomy, therefore, must reflect the lifestyle of the species' members, that is to say, the world in which they live.

I have said that the species society has no division of labor and is by itself a systematically incomplete and undeveloped society, but on the other hand it has the potential to develop indefinitely. The synusia is already a developed form of the species society, and the species societies that constitute a synusia have differentiated to become parts of the whole. Thus we might say that a division of labor can be recognized, but it is only a two dimensional elaboration of the species societies. The synusia, like the species society, is systematically incomplete. However, in the case of the synusial complex a division of labor exists to a considerable extent. In the previous chapter, I explained this as a division of labor like that between predator and prey, but I did not limit this to the relationship between those two categories. If organisms live in the same place, but eat different food, it is a division of labor. Even if they eat the same food, but the method of obtaining the food differs, then it is also a division of labor. Considered in this way, the synusia as a phylogenetic community may be compared metaphorically to a community of members that has the same job, or a workers' guild. By contrast, in a synusial complex there are various members with different vocations. In that sense, as members of various vocations gather and live communally, the synusial complex can be said to be a socially more developed community than the species society or synusia which are simple phylogenetic communities. Of course, the synusial complex can be thought to have its own wholeness. However, in the synusial complex where those who have different kin relationships and vocations live together, there is no reason that each living thing must share the same destiny. Each tries to live autonomously to the end. Hence, the wholeness of the synusial complex can be seen as simply an equilibrium which is recognized in the coexistence of these different members. Suppose we recognize this condition of equilibrium as a structure of this society; then its wholeness is wholeness based on this structure; as long as we call it a structure, we can expect a kind of completeness in it. In the synusial complex, we think for the first time of social organization.

Even the synusial complex is not complete in itself. It can be thought to correspond to one class in a social organization. In a single class, various vocations can be seen. There are honest hard workers, influential persons, and nonproductive parasites of the society; but the class is a class because of the existence of other classes. Therefore, if the coexistence of several synusial complexes is viewed as several classes existing together, then here, for the first time, the community of living things as a whole community can be recognized as a fully integrated social organization. That is, we recognize its completeness, not only as a whole community of living things, but as a social organization. However, if we are to debate seriously on the completeness of this whole community, it is not enough to look at just any community at random. The fact that a randomly chosen community of living things is a typical whole community indicates that it is one realized phase of a community of living things in a given place on earth. The expression of that phase originates in the habitat segregation of species in various synusiae. Unless this habitat segregation takes place in exactly the same way in every synusia, a typical community of living things does not show the morphological completeness that is seen in a phylogenetic community. This also applies to the synusial complex. Therefore, the coexistence of each member in what is seen as a territorial community, is really a kind of coincidental encounter at a particular place. Although the principle by which they are integrated is based on an organizational equilibrium, if this balance is achieved, to put it in an extreme way, such things as the affinity or characteristics of the members are not very relevant. However, if we consider affinity or qualities, or the phylogeny of living things, or the typology based on this species composition, we have to go back to the synusia which I described as two-dimensional or nonhierarchical, and follow it to its final divergence. The limit to this divergence ultimately must be the earth's limit. Thus, the real completeness of the whole community of living things has to originate from the completeness embodied in the earth. And only when we think of the living things of the whole earth, can we think of the real wholeness of the society of living things. Thus, the one community of living things which contains all the living things on earth, is the world of living things as a whole so to speak. We can thus understand the world of living things as having the individual and the world at opposite extremes. Each individual is at the center of its own world and is connected to the whole world through the species society, synusia, synusial complex and whole community. Of course, each of these societies can be thought as a center of the world. However, by understanding each of these societies as a place in which an individual is located, we can say that the individual which is one extreme of the world always is attached to the world and always is interacting with the world.

The world of living things or this community of living things as a whole society has a self-completeness in the sense mentioned, which means that this kind of whole community will not continue indefinitely. One whole community starts with its own features and develops by utilizing them to the extreme. Once it reaches the summit of its development, sooner or later it begins to self-destruct and by its collapse, another whole community with different features begins to develop. Although the various organisms originally divided and developed from one thing, it would be a mistake to conceive of the history of this world of living things as one species' linear development like the gradual continuous growth of our body. If we look at the palaeontological fossil record, it is immediately apparent that the so-called whole community has encountered radical changes and been rebuilt anew many times. The Mesozoic era is called the age of reptiles because throughout it, apart from the very earliest times, the reptiles, which constituted a phylogenetic community, dominated the world of living things and occupied the ruling class of the age. It is not that the other animals did not exist during that era. Both fish and insects were already there. However, in that era the development of reptilian features was the development of the world of living things. In other words, the reptiles created a new world, while the fish and insects did not participate actively in the historical development of this period. We do not know the details of how the reptiles rather than the fish or insects came to occupy the ruling class. But by occupying this position the reptiles' big step in creation, which is the period's main feature, became possible. Then as the dominant synusial complex in the whole community, there were no other organisms to suppress its members. In this way, the reptiles adapted to many kinds of environments, extending their synusiae and differentiating into synusial complexes. Meanwhile, predators and prey vied in size to produce ultimately monstrously huge reptiles. We can imagine that the world appeared as totally the kingdom of reptiles.

However, this reptile kingdom ultimately collapsed. Some think this was due to climatic change, others attribute it to a change in the earth's crust. Thus, for example, geologists explain the reptilian extinction as due to the so-called Laramide revolution. Although a drastic change in the environment probably helped to trigger the collapse of this great kingdom, I do not think it was the sole reason. The cause lay, instead, within the living organisms themselves. We have to consider that it was inherent in the self-integration of the whole community. The least likely scenario is that the reptiles lost out to the more intelligent mammals that subsequently became dominant. That reflects the thinking of those who do not know the structure or social organization of the world of living things. I said that synusial complexes are classes in the society of living things, which are in a relationship characterized by their mutual disjunction. In that sense this is a kind of hierarchical society. It is a limited and conservative society that cannot compete randomly with and replace other classes. It maintains its own position in the social organization, thereby maintaining the equilibrium of the society of living things. There also the integrity of the whole community is recognized. When the dominant class of this society collapses, the society does not weaken; instead, something new appears and establishes a ruling class that replaces the former. What can this mean? Here, we cannot help but recognize the self-containment of the social organization of the whole community of living things, while the synusial complex is a class that is not yet independently integrated. We can compare the action in which this whole community creates anew the lost class with that in which the individual organism, which has its own wholeness, recreates a lost part. If we see in these regenerations a kind of congruency and recognize the similarity of autonomies based on the wholeness of both, such autonomy also probably exists in the species society, synusia and synusial complex. However, in the species society where we do not see specialization, regeneration is achieved simply by individual reproduction. In the case of the synusia or synusial complex, the formation of a new species is required. Within the whole community, several quite new species must be formed in the regeneration of classes. Therefore, it is no longer simply regeneration, but creation. Regeneration in the whole community must be the greatest embodiment of creativity in the world of living things.

We can see this most vividly in the evolution of the mammals and birds which became representative of the Cenozoic era after the extinction of the Mesozoic reptiles. Mammals, for instance, adapted to occupy the empty life-fields, the empty places, so to speak, left by the now extinct reptiles. Eventually, almost without exception, these places were filled by various species of mammals. Herbivores and their predators emerged; marine, marsh, and tree dwellers; even flying mammals such as bats. Further, the social status formerly occupied by the huge reptiles was inherited by the emergent huge mammals. Thus, with the mammals established in the once defunct ruling class, and the whole community of living things having regained its equilibrium and completeness, the Cenozoic appropriately came to be called the age of mammals as befit their place in the social structure. The peak of mammalian development was probably when the elephant became the most powerful terrestrial animal by virtue of its size, and its race was the most widely distributed on the earth. Most of that race of elephants is by now extinct and some of the descendants have been domesticated. Even understanding that everything must change, we cannot help but feel regret to see a domesticated elephant burdened with loads, and recollect that the elephant has already passed the peak of the age of mammals.

I stated previously that although we do not know the process by which the reptiles became the ruling class, we can assume that they made the creation that represented that era and that neither fish nor insects actively participated in that creation. This applies likewise to the age of mammals. Not all reptiles became extinct, and crocodiles, snakes and turtles exist today. However, in the period of the spectacular creation and evolution of mammals, how much did these survivors evolve? They must also have had to change to adapt to the new period. They have survived until now because they accomplished that change, which is none other than evolution. More than evolution, it was a creation from something. In general, in this spatial and temporal world, absolute stasis is not possible in anything. The very foundation of life in living things lies in the impossibility of stasis. For living organisms, living means acting, and those which are created will create those which in their turn will create again. Both in the growth of the individual and in the continuation of generations, theoretically there is no simple repetition, but somewhere new things are always created. We cannot doubt that evolution is creation and that creativity is an attribute of living organisms. By comparison with the spectacular development of mammals, the surviving reptiles have evolved slowly, almost incomparably slowly. Evolution does not occur at the same rate in all living things. To live is to act and to create. In that sense all the daily life of living things is part of evolution. General evolution is in reality minute as shown by the surviving reptiles. It is so slow that even after several hundred generations and several thousand years, we can hardly recognize any change. Of course, evolution is not seen within our short lifespan. This applies not only to reptiles. Throughout the period of mammals neither fish nor insects displayed any notable change worth calling evolution. They scarcely evolved, then, throughout both periods of reptiles and mammals. They were a ruled class, or, so to speak, the common class. It was irrelevant to them whether the rulers were reptiles or mammals. Only the ruling class evolves sufficiently to represent an era. In the Cenozoic, because mammals replaced the reptiles as the new ruling class, we think that only mammals achieved a phenomenal evolution. However much the common classes accumulated minute changes in their daily lives, these could not come close to the rapid evolution of the mammals.

But from where did the mammals, which leapt into such great creativity to become the favorite child of the age, originally emerge? In what corner of the society did they hide during the age of the reptiles? And why could they, over other animals, succeed the reptiles as the ruling class? The fossil data have not yet provided clear answers to these questions. Therefore, I would like to rephrase the question: When the reptiles collapsed, why did the insects not replace them as the ruling class? It is thought that even insects may once have occupied the ruling class, based on the existence of a huge fossil dragonfly with a wingspan of nearly a foot. But, naturally, even if there was a period when insects ruled, it must have been a much simpler social organization than the period after the more advanced reptiles and mammals emerged. Meanwhile, early vertebrates emerged from the water, expanded their territory and invaded the land. They probably evolved into reptiles from a stage where they lived partly in the water and partly on land much as modern amphibians. What kind of changes were wrought in the social organization at the time by the emergence of this unanticipated invader from an unexpected place? One way to avoid mutual destruction and to coexist could have been to form synusiae or a synusial complex, but these two phylogenetic communities' lineages were too distant to do so. The only remaining solution probably was for these two phylogenetic communities to each form by its respective classes separate synusial complexes. The result was the addition of another synusial complex and the elaboration of the existent social organization. We can suppose that subsequently, insects relinquished the ruling position to the amphibians and gradually reduced their body size, while the amphibians, as the ruling class, gradually increased theirs. Although this is only conjecture, probably the reptiles followed the same path as the amphibians. This is expected from the phylogenetic relationship between reptiles and amphibians. And indeed, contemporary birds and mammals have the closest affinity to the reptiles. Although in each evolution a different ruling class emerged, this was only a change of ruling class within the same single vertebrate community. At least since the vertebrates emerged, the ruling class has been monopolized by them. Viewed in this way, although the period of reptiles itself probably had its self-completeness, we cannot think that the age of reptiles and the age of mammals were completely separate in terms of affinity. Rather, the ancestor of the mammals, which built the age of mammals, did not suddenly appear, but already existed in the society of reptiles during the age of reptiles. Since we see that the reptilian society through a revolution changed into the mammalian society, we also think there was an extinction there. However, if we consider that through this revolution some reptiles metamorphized into mammals, we see the continuity. As mentioned, the social organization of the mammalian era is not an elaboration of the reptilian era's organization, but is more like a reproduction of it. Thus we can also say that the mammals, which were actually metamorphosed and evolved reptiles, reproduced the age of reptiles. In the fact that the next ruling class comes from the ruling class which is destined to collapse, we see that the classes in the society of living things carry to the end the characteristics of a hierarchical society. Further, because it is a class society, and insects are a class of the masses, insects could not, after all, have become the ruling class.

Once the insects had yielded their ruling class status to other animals their future course was inevitable. In order to avoid conflict with other synusial complexes which occupied dominant positions they became smaller and smaller seeking and settling in places where larger animals could not invade. However, there is something we must remember. Organisms have only one body as the instrument and means by which to live. Moreover, that body is inherited from their parents and within it the history of everything that was experienced by their ancestors is inscribed. Just as we cannot change the past, we cannot do anything about our body. It is already made, a given thing. Those without wings cannot fly. Those without fins cannot have a diving life in the water. That is life's limitation and restriction. Therefore, if we liken their mode of living to a function of a livelihood dependent on the body, we can say that organisms have hereditary living skills. But of course the present body did not exist from the beginning. The fact that it was made means that something made it. Insects also must have had a prototype which expanded into various environments at some period. Now, the extension of an organism's autonomy over the environment has meant reciprocally the environmentalization of the subject as well as the body. Even if organisms had a freely creative potential within themselves, the environment restricted it. In that sense we can say that the environment, which was itself created, conversely also created the organisms dwelling therein. And because the organisms that were made in this way also sought that kind of environment as the most suitable one, as this interaction progressed, or the more the body specialized, the more difficult it became for the organisms to leave that environment and the more they became committed to that course. Even so, organisms had not lost their creativity.

I think that they had their hereditary vocational skills and protected their working place and gradually transformed these natural skills into a specialization. That is the great expression of their creativity. But those which inherited six legs could do no other than make the best practical use of them. As a result, although contemporary insects are certainly smaller than earlier forms, their bodies are compactly efficient. The fossil-like, large-bodied forms have decreased and give a somewhat flabby impression. We can liken this to man-made machines where the more sophisticated and efficient they become, the more compact they become and can even be said to have a kind of beauty. However, for the same reason that they already had certain inherited skills, their society could only become a ranked society. The fact that they could not abandon their working place and skills and become the ruling class the next day, having already completely invested their body, which was their only capital, in the environment, must have caused great regret. Why, then, did they not contrive to change their direction before they reached that point? I believe they had enough creativity to do so. However, they were fundamentally pacific conservatists who liked the present state and avoided unnecessary conflicts. It was enough to maintain the condition in which they found themselves, with the maximum number of members having peaceful, communal lives, and making that communal living prosperous. Of course, this is an anthropomorphic explanation. In the order and equilibrium of the social organization of living things we can recognize the wholeness and autonomy of the community of living things; however, this has been the result of a development of social organization and its members have gradually increased. Social organization became possible due to the division of labor; namely, the division of skills and working places. The division into hierarchical, interrelated classes in the society allowed the division of labor to proceed. Thus, the world of living things was always one whole entity but it was also an integrated body of the world whose center was each living thing. Reflecting on the manifestation of the self-identical character of the world of organisms, which is like the notion of the one is many and many is one, soon becomes a teleological, anthropomorphic interpretation.

Theoretically, every individual, every species or class must touch the world and participate in its creation through this self-identical character of the world of living things. However, even limiting it to epoch-making creation, the participants in this creation are determined already by the division of labor. It is a privilege permitted only to those in the ruling class. Thus, the history of the world and of the evolution of living things may be defined as the history of the rise and fall of the ruling class. Moreover, this ruling class is contained in a class society and is born from the ruling class. This is fundamentally the way of living things. The history after the Mesozoic is, in short, the history of the development of the vertebrate community and of its rise and fall as the ruling class. Mankind arose within the community of mammals, and has replaced other mammals temporarily as the ruling class of the society of living things. The history after that is the real history of mankind. Those with a superficial knowledge of evolutionary history tend to speculate on what would rule the world after humans and suggest that it would be bacteria or insects. However, the notion that the descendants of bacteria or insects would become the ruling class after humans is as absurd as saying that the descendants of humans would become something like bacteria or insects. These things will not occur. Even a creative evolution that deserves to be called an important event in evolutionary history is only possible in the ruling class of that period because of the division of labor. The new ruling class often makes the survivors of the previous ruling class one grade lower than themselves, and constructs a new social organization. Although the living things that constitute the contemporary world all live in the present, they differ from one another historically and epochally. If these different things lived together in confusion, we could not say there was any order in the social organization of living things; but it is very interesting that they are hierarchically interrelated in intermittent continuity, each as a class. In this organization, as long as they occupy their proper rank, even those of the past period can continue to exist, and at the same time, those that wish to survive must devise a way to protect their position. Considered this way, each living thing has only one opportunity to play an historical role such as an epochal evolution. At present, mankind occupies the ruling class and is in the phase of creative evolution; all other organisms have already played the role when they had their chance. Of course, existing bacteria or insects are not the same as they were during their golden period. However, in this sense, between bacteria and insects or even the more recent mammals or birds, there is no difference because they have already played the role. It is often said that nature is periodic or repetitive, but that is only nature in the abstract. The nature surrounding us is not like that. When we look at living things in nature, at the very least we cannot say that evolution is simple repetition. In living things in general although we cannot recognize the same kind of individuality as we see in individual humans, each species of organism has its own individuality and species history. Does this not cast doubt on the idea that nature is invariable and that history exists only for humans?

To return to the earlier question: what will rule the world after mankind? The current rule will probably continue for the time being, but human development has its own limit. This should not worry us, however. Those who replace us, though they perhaps should no longer be called mankind, will originate and be created from the human race. This is what evolutionary history teaches.

In this small book I have no intention to describe the origin of the species or the mechanism by which a new species is produced. That is a problem addressed by contemporary experimental genetics, and a field in which an amateur such as myself is not competent. However, since Darwin, the origin of species has come to be regarded as the central issue of evolutionary theory. Where it is relevant, I would like to reexamine what I have already said from the point of view of the origin of species, and carefully scrutinize on what points the origin of species can be considered as the central problem of evolution. If possible, by making clear the fundamental differences between the evolutionary theory on which my world view is based, and the current orthodox view, I would like to conclude this short book.

As previously stated, the period in which any particular phylogenetic community achieved a sudden burst of evolution was historically determined.

In that sense, the present is indeed the age of mankind. Organisms that previously achieved tremendous creative evolution now scarcely seem to be evolving. They have occupied a certain rank and have certain skills in the social organization of all living things. To state it differently, through that kind of hierarchical relationship or relationship of vocational skills, they can live communally with other members of the community of living things. An indiscriminate change in their class and skills would destroy the social organization. The body inherited from their parents is the concrete expression of their class or skills and cannot be changed in a short time. They do not unnecessarily engage in wasteful conflict and thus the social organization achieves its equilibrium. Simply stated, only those who occupy the ruling class, because nothing exists to dominate them, can to a certain degree continue their evolution. My opinion can be summarized thus: I do not deny at all the creativity of organisms which in the past achieved spectacular evolution but in which we now see almost no evolution. All organisms are creative, but whether evolution is encouraged or not ultimately depends upon their social place in the whole community of living things. Of course, we cannot say that this creativity is the same in all organisms. However, when humans domesticate wild animals several varieties emerge. What causes this variation? In my view it is because those animals have been deprived of their membership in a natural community and transferred into a human society. Under human protection they become domesticated and are given a totally different living place. In other words, the social restrictions on evolution in a natural community are lifted, allowing this creativity to emerge. The fact that Darwin, extrapolating from the variations he saw in domestic animals and plants, thought that variations exist in the same way among organisms living in natural conditions, makes me suspect that the concept of his evolutionary theory may have been wrong at the beginning.

Needless to say, the variation that is the issue in evolutionary theory is genetic mutation. No one would deny the general argument that these kinds of mutations accumulated through innumerable generations over thousands, tens of thousands or even millions of years and that even those plants and animals leading natural lives gradually change. To deny this is to deny evolution. However, is this variation really like the random, capricious variation Darwin recognized in domestic animals? From among these random variations humans kept what they wanted and eliminated what they did not. This is artificial selection. In the same way, through the sieve of the struggle for survival under natural conditions, those random variations which were suited survived, while the others could not. Thus, gradually, only the descendants of the adapted ones prospered. This is the theory of natural selection, which was suggested originally by artificial selection. Although it simply replaces human agency with nature, it has held sway over a generation as the theory of evolution. However, are the variations in nature indiscriminate and random, or can they be? Of course, this randomness does not mean that a fish can be born of a human or vice versa. An eggplant does not develop from the vine of a cucumber. The body of the parent as created in the past ultimately limits the body of the future child. Within these limits the range of all possible variation is nondirectional.

The body, however, is also an expression of a lifestyle. The organism and the environment influence each other through the body and herein lies the foundation of living. In that sense, if we think of the body as autonomous to the organism, it may be accepted as a thoroughly independent thing; but if we consider it environmentally and materially, the body is ultimately an extension of the environment and nothing more than one representation of the environment. Now, is it not the case that the theory of natural selection never admits the reaction of organisms to the environment, but only the influence of the environment over living things? I say this because if organisms show random variations in the sense stated above, then they cannot recognize the environment and they have a completely blind existence. If that were so, then even if we did not use difficult concepts such as "autonomy," we naturally would not understand the meaning of the living organism. To live is the opposite of dying. When we say that organisms live, between the options of living or dying, we can say they live because they chose to live. They chose to live because the fundamental principle of the existence of living things is guided by the basic principle of the existence of this world. Therefore, it is perhaps natural that living things opt to live, but behind that inevitability, cannot the freedom of choice faintly be felt somewhere? For example, living necessitates eating food, avoiding enemies and seeking a mate, but food, enemies and mates are all a kind of environment. Therefore, in recognizing these things, organisms specifically choose them from the whole environment. That is to say, to recognize is to choose. It is not that something becomes food after being ingested, or becomes prey after being eaten, or becomes a mate after copulation. Only when living things react, does the environment make them live. If organisms do not react, the environment presumably kills them and transforms them into matter. The recognition of the environment is the organism's reaction to it; it is the choice of living things in the environment. If we see that as natural, it is instinctive. But even if it is an inevitability in nature, there is a freedom within this inevitability. It is nondetermination within determination. The evolution of living things becomes an unsolvable mystery if we see it as mere inevitability or determinism.

Of course, variation in living things is the wellspring of evolution and the expression of creativity. It might be said, therefore, that the theory of natural selection assumes random variation and fully admits the creativity of living things. But if, ultimately, only those superior individuals remain which are selected by the environment, or the so-called survival of the fittest, then living things are not creating, but are gambling. Evolution is not brought about by the freedom of certainty, but originates in the constraints of chance. Since the appearance of living things in this world, how many millions and billions of years have passed? Through this time all living organisms have survived by reacting to and being influenced by the environment. As for variation in living things, is it possible that they separated themselves from the guiding principle of their lives and existed for all these years detached from the environment, passively depending on chance? Even if the mechanism of variation depends on the subtle workings of the reproductive cells, these cells do not exist alone outside the body. If we think that the environment is an extension of the body, which is at the same time an extension of the environment, how can we imagine in the lives of organisms something that is detached from their body, and a part that is indifferent to the environment? This is characteristic of the period of mechanistic thinking, in which organisms and the environment were thought of separately, and abstract living things and an abstract environment were held to be connected in a causal relationship. But that does not begin to explain the way organisms actually live.

I said that the character of the life of organisms is the assimilation of the environment through the body, and conversely, the environmentalization of the subject through the body; but as I described before, the body is both free and not free. Through the conflict between this freedom and destiny, a new body is created. Can this not be understood as variation? Let us say for now that the evolution of living things, which inherit and bequeath their body, is, in this way, the creation of the body. I do not think that in the lives of other organisms there is necessarily purpose such as in ours. In the choice of whether to live or to die, living things are beings which choose to live; only in that fact is the living of organisms already directed. The drive to live is firmly established. How, then, can living things, in which the subject and environment dynamically interact, exchange the potential to create the body for gambling on its fate? Variation itself must be the environmentalization of the subject and the extension of the subject's autonomy over the environment. It must be one expression of living, or rather, the expression of living well. When the maintenance of homeostasis means death, then the organism attempts to live better in whatever possible way. Because the life of organisms is thus directed, the environmentalized subject tries all the more to internalize the environment and is all the more environmentalized. So the principle of adaptation probably lies here. Such notions as random variation are abstract products of considering living things apart from their way of living. Living organisms follow a certain course of living. That course is neither determined by the organisms nor by the environment. It is the directedness of creativity determined by the freedom of necessity. Even so, I do not intend to assert that acquired characters must be inherited. Because the vitality, adaptability and creativity in the body of the parents cannot continue forever, it changes into the body of the offspring. It is enough if the body of the offspring is equipped with the variation which is suitable for a better life. It is also not necessarily limited to particular offspring of particular parents. It is sufficient if the adaptive potential increases in each individual within the species as a whole.

Here, the origin of species also becomes an issue. Because random variation was assumed, it was necessary to postulate natural selection. This combined assumption of random variation and natural selection arose from thinking only of the effect of the environment on the individual and disregarding the influence of the subject over its environment. But where this dynamic between the environment and the subject exists equally from the beginning, completely random variation cannot exist. From the beginning variation is guided by the directedness of each species' way of living. Does not that, then, limit the field for natural selection? By directedness, I do not mean a linear directionality. I mean that instead of 360 degrees of variation, I think of a 20 or 30 degree angle and regard it as directedness as opposed to complete randomness. If the fact that old individuals die, and that the sick, weak, and injured cannot maintain themselves is regarded as natural selection, we cannot deny that process. But if that is natural selection, it is not a convincing explanation for the origin of species. In the first place, what is the species? It is the territorial extension of individuals belonging to a kinship community, which, by virtue of having the same morphology, have the same way of living, and in turn through following the same lifestyle, have similar bodies. If we take this a step further, we can say that because they subsist in the same way, they have the same direction of living and accordingly will tend to express the same variation. All individuals do not have to show the variation at the same time. In succeeding generations, there is a gradual increase in individuals that display that kind of variation; at some point, the species itself changes. The general conception of the theory of natural selection is that changes in a male and female pair are expressed in their descendants, which become the winners of the struggle for survival, gradually destroying and replacing the descendants of those who do not change, and expanding their territory. However, the variation tendencies of the species itself is already determined. When the species itself changes, those individuals that changed sooner than the others are, so to speak, the forerunners, having simply matured earlier than other members which, sooner or later, also change. The origin of a species can be compared to the origin of an individual: there emerges a new species which did not exist previously. On the other hand, to think that like the individual, ancient species that existed before have to die is an easy mistake to fall into. If we think that a species gradually changes in a certain direction, this transformation should be compared to the development of an individual; even though the larva, pupa and adult look very dissimilar, as long as it is a difference in the phases of development of one individual, we regard it as an identical individual. Likewise, during the process of changing in this way, the species itself becomes very different from what it was like at first. Though palaeontology may judge the forms at either end to be separate species, they constitute one continuous sequence. Should this also not be regarded as the development of one self-identical species?

For the species then, as for the individual, does there come a time to die? I have commented not on the origin, but on the death of species, because where there is death, there is birth. I think that a proper understanding of the death of the species will at the same time provide an insight into the origin of the species. Certainly, many species of living things have disappeared. Without doubt, there were species that left no offspring. But if species, like the body of an individual, must perish, then no living things would be seen on earth today. In the sense that they all developed ultimately from one thing, contemporary plants, amoebae and animals must have lived through the same periods and epochs on the earth. In this sense we can say they have lived a tremendously long time, and even that they do not perish. However, although they are the same animals, relatively speaking mammals are newer than the reptiles. By new I mean that it was new that they emerged as mammals on the earth, and mammals are therefore younger than reptiles. I said that through the great revolution of the reptilian extinction mammals were born and that reptiles transformed into mammals. That is to say, the origin of the mammals is inherent in the origin of the mammalian species.

We do not know much about the ancestors of the mammals that had the heavy burden of rebuilding the ruling class after the reptilian extinction. But those ancestors showed spectacular creativity. Speaking anthropomorphically, they acted with absolute determination. Indeed, on that occasion it can be imagined that all possible kinds of variation emerged. We should note that on this occasion this variation experienced little natural selection and proceeded rapidly; or conversely, because it was a time when natural selection had little effect, living things expressed a very broad variation. Here I would emphasize the inseparable relationship between living things and the environment. Although organisms generally possess a directionality that shows only modest variation except in the above kind of case, we cannot deny that they are capable of great mutability. Thus, in these two cases, the one in which variation is directed corresponds to what I previously called evolution in daily life; the other example of nondirectional broad variation corresponds to the organizational change of one whole society. These can be called microevolution and macroevolution respectively, but in neither case is natural selection seen to have the effect predicted by that theory. As previously described, while we think of microevolution as the development of the newly born species, the origin, that is, the creation of a species is linked with the extinction of the species. Only by relating the origin of the species to the macroevolution which accompanies the metamorphosis of the social organization, can we perhaps find the proper foundation for discussion.

From the beginning it has not been my intention to pick out the various problems of evolution and criticize them point by point. That would require another book. I have only, from my point of view, declared my disagreement with natural selection theory, which is regarded as the orthodox theory of evolution. However, it may have been very hasty and ill-prepared. By attempting to sketch the main points, some points may have raised doubts in the reader. For example, when I say the interaction between the organism and the environment, it may be understood that all living things have to adapt themselves perfectly to their environment. However, it is not uncommon to find organisms which are not perfectly adapted. Even in mankind, who became bipedal long ago, it is said that detailed examination of the leg bones, for instance, reveals we are not yet perfectly adapted to an erect posture. Yet it is a basic principle of life that the body always is led toward adaptation. Thus, it also can be thought that adaptation commences first in the parts of the body that are mandatory for survival. What taxonomists select as the distinctive features of a species is often very small, such as the markings on butterfly wings, or the arrangement of hairs on the abdominal segments of a larva; but the presence of one or two spots on the wings, or a slight difference in the arrangement of hairs cannot be thought to directly affect their way of living. In general, microevolutionary changes often occur in characteristics that are thought not to be directly useful for their survival, from our point of view. Conversely, this can be understood as organisms change because the modifications present no problems for their way of life. Since they are members of the same species with the same way of life, the direction of change becomes the same; even so, there can be differences in the degree of change among individuals. These differences simply reflect the diversity among the individuals. But this range of change among individuals does not indicate the absolute limit of mutability. Living things with the potential for random variation always have a limit relative to the environment. Yet there may be times when mutants go beyond this limit. The problem is what is the destiny of these mutants.

In general, the fact that such mutants are weak is already confirmed in my view. But if we supposed for a moment that they were vigorous, since the parts that change do not have a direct influence on their living, natural selection, which anticipates the survival of the fittest as its consequence, has no selective power. This again contradicts the theory of natural selection, but I do not deny that, very rarely, mutants do appear in the natural world. I simply think that the emergence of this kind of mutant and the origin of species should be considered as very different problems. Within the relative limits of their potential for variation, species in the natural world do not change indiscriminately. Rather, a species suppresses extreme forms and by thus modulating variability maintains a state of equilibrium. This also can be seen as an expression of the tendency to preserve the status quo or the conservatism of living things. But this maintenance of the status quo is, after all, the maintenance of the social organization of the world of living things, which in its turn is the sustainment of the structure or system of the world. Therefore, leveling the ground of mutability is, so to speak, dependent on the mixing of the genetic substance and prevention of the emergence of an unstable and weak pure line. It is also a concentration of variation to make as many individuals as possible maintain a middle path of change and this can also be thought of as the strengthening of the species itself which is probably one expression of its autonomy. Therefore, within the possible domain of mutability of a species, there must be a mean point of variation around which numerous individuals of that species cluster. But if we interpret it statistically, the variation expressed by most individuals, or something like the average variation, may be considered the equivalent of not changing at all. But statistics reports on static things; because it sees species as unchanging, it generates those results, but even species are constantly changing and growing. If we accept this, that kind of change in a species perhaps should not be understood as the change in the mean point of the species' variation as described above. The movement of that mean point must indicate the direction of change that the species will take. Thus, what does the trace of this mean point show? Of course, it cannot be a straight line. It also cannot be a level or two-dimensional thing. Variation or evolution, which cannot be considered at all without including the time factor, cannot be two-dimensional. At the moment when the mean point moves either right or left, we can consider it as being two-dimensional, but its historical track is most probably a spiral. Although it presents a spiral in the long run, the mean point of variation moves either right or left at a particular moment and the immediate problem of whether to move right or left is already decided. For only when that is determined, can we contend the directedness of variation. If so, the fate of a mutant which has deviated from these directional bounds of variation may perhaps be likened to that of the pioneer who, foreseeing the future prematurely, is trampled on by the unthinking mass. In the case of humans, such unlucky pioneers can be recorded on a page of history. But for living things, I believe that the history of the species is the history of the living things. This also means that the so-called mutation expressed by individuals cannot become materials for tracing evolutionary history.

There is no need to say that this applies in instances where we consider a particular characteristic as a unit of change. Of course, in the body of a living thing certain characteristics actively change in a certain direction, but there are other characteristics that are conservative and offset the active change. Furthermore, change does not have to be limited to the structural and morphological aspects of organisms; the same can surely be assumed of their functional and behavioral aspects. The problem ultimately is always that the morphological characteristics used by taxonomists to define each species, such as the markings on the wings of the butterflies, or the number of hooks on the penis, are not correlated with a characteristic function or behavior. In addition, many of these characteristics cannot be considered indispensable for living and therefore cannot be thought to have been acquired by following the organism's guiding principle of survival. Or, this may be due to our ignorance. But we must begin to wonder why characteristics that we think are nonessential for survival, at least so far as we presently know, developed as special adaptations of the species. The earlier analogy with the specialization of vocational skills may be relevant here. That is, when the habitat is fixed, and the way of living is shaped to fit it, and the body also adapts materially to this, we say it has specialized. This can be called, for the organism, the culmination of adaptation. It may be that organisms that have reached that state have no need of further bodily changes to meet the demands of their lifestyle. Or, perhaps they cannot change; awkward remodeling would only destroy the hard-won splendid adaptation. As long as their continued existence remains possible, we must recognize the creativity even in these species that have reached an adaptive culmination, and that no longer express it in further adaptation to their lifestyle. Perhaps this creativity can only be revealed through changes in characteristics that have no direct bearing on the means of survival. People often think that the life of organisms is a lifelong, unceasing quest for food or sex and that they have no other life outside such drives. I also think that remaining alive is the principle that orients their lives and until now I have tried so far as possible to interpret living things from that perspective. But is that explanation sufficient to understand the totality of ways of life expressed by living things? Is it true that their lives begin and end in wretched brutishness? If living things are like that, then why are flowers and butterflies beautiful? To have to think incessantly of bread in fact runs counter to the world of living things and is a concern only for humans who have become estranged from this world; as long as all members of the community of living things in nature are content with their status and remain in their working place, if we consider only their subsistence, are not their livelihoods far more secure than that of humans? Of course there are enemies and disease in their world. But, rather than only the negative aspects of living, can we not think more of the positive aspects? I would not disagree that both improved adaptation and the development of social organization came from the fact that the better life sought by organisms originated in a secure subsistence and avoidance of useless conflicts. But what was better living for organisms whose life was already secure? Are living things indolent; do they, while eating or sleeping without expending any effort, unknowingly become beautiful? Of course I do not think that organisms understand beauty as we do. But I would frankly admit that there is an aspect in living things or in the life of living things that cannot be explained only in terms of a drive to survive. That is, intentionally or not, living things gradually became beautiful. For instance, an often cited example is that of the ammonites, which lived in the Mesozoic seas. Over a very long time during which the species grew, the engraving on their shells gradually became more refined and delicate. Is this not something like art in the world of living things? And is there not something that could be called culture, although it is of course different from human culture?

Why, in fact, do taxonomists choose the distinctive features which, on the whole, have no direct bearing on their means of surviving, to distinguish between very similar species? Closely related species are fellow kind which constitute a synusia as one phylogenetic community, and these are the most similar in terms of affinity. Because the formation of this kind of synusial species was understood as an adaptation to the living niche of one phylogenetic community, the limiting effect of this niche on each species must be unmistakable. In other words, the influence of the environment should be manifest where it had a direct relationship with the species' way of living. However, this may also be an inevitable result of our present ignorance and a closer inspection might reveal such differences in unexpected insignificant places. But more than that, if there is a clear difference that anyone can see readily in a feature that has no direct relationship to the means of living, then what after all becomes the distinctive feature of the species is its cultural characteristics, and taxonomists distinguish between species by

82   On history

these cultural characteristics. Then the origin of the species is a problem of how did these kinds of cultural characteristics diverge and develop to the point where they became a defining feature of the species. Of course, even a so-called cultural characteristic must be regarded as a variation which evolved. Its change must also be directional, as in the case of the ammonite shell that gradually became more finely engraved. However, as there is no adequate general guiding principle of living underlying the independent development of these cultural characteristics into distinguishing features of species, as a final recourse we must think of the species' innate predisposition. But what is meant by the species' innate predisposition or ability?

The species' innate predisposition is the innate ability of the blood. To express it in nonmysterious or scientific terms, it is the genetic predisposition of the species. As mentioned previously, the species has a tendency to concentrate the variation expressed by individuals by making their genetic constitution as homogeneous as possible. Also this tendency is understood to be an expression of a species' autonomy. Therefore, the conflict between species is the conflict between the autonomy of each species. I suggested that if the entire environmental surface of the earth were uniform, one species might have spread over the earth. However, since the surface of the earth varies, the different species that adapted to different places of living separated as synusial species. These synusial species, instead of a single species society, extended as phylogenetic communities of synusiae over the earth. I also suggested that synusial species form as a means of maintaining equilibrium by separating individual groups with different tendencies. What these tendencies mean, in sum, is that the difference in tendencies is the difference in predispositions, and the difference in predispositions is the difference in tendencies. Although I refer to the difference in predispositions or tendencies, the problem here is that if this takes the form of the rarely seen mutant, it likely cannot act contrary to the autonomy of the species and separate from it. That is because a separation is the division of the species' autonomy, and thus has to originate in a conflict in the autonomy. Therefore, it is logical to think that underlying the differences in individual predispositions which lead to this kind of separation, there is commonly an environmental influence. Whether that happens or not, if there are individuals of a species that prefer not to dwell at the center of the distribution where it is crowded and difficult to live, but prefer a peripheral area even if they have to deal with a somewhat difficult climate, the individuals in the peripheral and central areas not only have different tendencies, but their differences are likely related to the difference in climate. Indeed, we may think that as with the distance that separates the central and peripheral areas, their kin relationship likewise is distanced. But the separation is not achieved only by this. It requires the separation of the autonomy and the independence of the species.

At present, the process of separation is not entirely clear, but its conclusion, namely, the independence of the species, is achieved when the number of individuals that do not clearly belong to either species gradually declines and interbreeding largely ceases between them, even though they border on each other. Therefore, the separation of species can be thought to mean the attainment by two species of their own autonomy. However, we cannot conclude directly from this that upon reaching that stage, two species are separated in terms of affinity to the extent that distinctive features can be recognized between them. I would like simply to emphasize that this separation is an indispensable condition for the formation of the species. A species is a species because it does not interbreed and maintains the integrity of its constitution. Therefore, we can assume that individuals are able by some means to distinguish between individuals which belong to their own species and those which do not. Although there are examples of interspecific breeding in plants, and cases have also been reported in animals, we cannot use these rare exceptions to argue the general case. Although some environmental influence underlies the separation of species, there are many cases in which its influence is not necessarily expressed in the external form, but can be seen in physiological or behavioral change in the subject. We can also say that even those that have separated, but are phylogenetically close, form a synusia because they have the same fundamental life forms. Their fundamental characters and basic propensities are the same. The changes gradually generated in nonessential cultural features, which have no direct relation to their way of surviving, originate in the characteristic differences, which though negligible at the time of separation, unexpectedly develop into the distinctive traits distinguishing each species after separation. Unless the distinctive features that we recognize actually help individuals identify other members of their species, it may be better to assume some chance factor here. It is not absolutely necessary to use a word like chance if it is misleading, but, for instance, to the question, why did science develop in Europe and not in Asia, many people attribute it to differences in climate. However, since a European type of climate can be found in both America and the southern hemisphere this relationship does not necessarily hold. It is illogical to assume that the confinement of scientific development to Europe originates in some chance factor that accompanied the separation of fundamental characters. Rather, such a chance is shared by many individuals, and where, through this concentration even on such cultural change, the species manifests its own autonomy, something random yet not random can be sensed. A species also can be thought to be directed by some unconscious design, so to speak. We can even think of it as the destiny of things that are endowed with creativity. Thus, I venture to borrow the term culture for the life of living things to best express the flower which blooms on this destiny.

Now if, for the formation of the taxonomically close species being discussed here, the separation of basic characters like this is essential, then in the case of terrestrial animals, when one continent divides into two, the traffic between them is obstructed by the sea and, inevitably, this separation has an effect. It is reasonable to think that if the separation lasts for a long time, what was originally a single species will ultimately become several separate species. I cannot support the nonevolutionary thesis that the distribution of water and land on the earth's surface has remained completely unchanged since ancient times, but even if the distribution of water and land, or the arrangement of the continents and oceans were roughly the same as now, if the sea level rose a little, or the ground sank a little, then even without digging such things as the Panama Canal, North and South America could be separated, and conversely, if the land rose a little, North America and Asia would be connected by a land bridge on the Bering Strait. Our country is called $\bar{O}yashima$[66] and since the dawn of history, it has been an island country, but it would be very easy for it to be connected with Asia if the Mamiya and Korea Straits disappeared. Throughout the unimaginably long history of the earth, this kind of dynamic change in distribution of land and water was repeated and each time the separation of species was impelled. On the one hand, because the migration of species was also restricted or encouraged, we cannot necessarily conclude anything about the origin and formation of a species only by looking at its current distribution. However, a very interesting example of precisely this is provided by the flora and fauna of Australia, which are substantially different from those of other continents, to the extent that they provide current evidence that Australia separated from other continents in antiquity and remained separate thereafter. To take mammals as an example, with the exception of species of bats and mice as well as humans, who are globally distributed, only the most primitive mammals, the monotremes and marsupials, are found there. People are often misled by the term "primitive" and imagine the mammalian community in Australia to have a low and simple social organization. But the mammals in Australia have the same origin as mammals on other continents. They are alike in that after the extinction of the dominant reptile

---

66 $\bar{O}yashima$ = the eight great islands. The reference is to Japan.

kingdom they were born to construct anew a ruling class. But when some of these mammalian ancestors entrusted with this task colonized Australia, Australia broke away from the other continents. Since that separation was the separation of their basic natures, the mammals in Australia remained on the level of marsupials while more advanced modern forms appeared on other continents. However, though the mammals isolated in Australia had to remain at the marsupial level in basic character and structure, within those limits they performed their function superbly. The kangaroo is not their only descendant; there are predators similar to wolves which prey on kangaroos, and bear-like omnivores, as well as arboreal animals. Compared with the mammalian communities on other continents, it has, in its own way, largely perfected its social organization. That is to say, outside Australia there is one world of mammals which can be thought of as a theatre of evolution, while the isolated Australia mammals formed one small but self-contained world. The residents of that small world knew nothing of what was happening in the larger world, but they accomplished their own characteristic development and eventually became the unique community of marsupials in Australia. It may compare unfavorably with other mammalian communities and appear unsophisticated, but as it is, it has its own completeness. It is not that other mammalian communities once evolved through the same stage as the community of marsupials now found in Australia, nor that the marsupial community will eventually evolve into something like these other mammalian communities. Each should be regarded as having been, so to speak, a destined community; as long as they remained separate, each became qualitatively different, or a similar type of mammalian community became dissimilar and these were destined to co-exist and to develop concurrently.

The independent integrity of the species shown by this development is what is actually meant by a species' so-called destiny, and it has been the propelling force behind all evolution. This integrity or self-completeness is expressed in such things as individual reproduction and the natural renewal of plant communities. Nevertheless, the self-integrity in evolution has been always the self-integrity of creation. Can we assume this self-completeness in the evolution of the individual, in the fact that each living thing in its own way, much like flowers and butterflies, became beautiful? The integrity of the species is seen in the fact that this evolution is not left to the individual's whim, but usually follows the whole species' evolutionary direction. As mentioned before, in the formation of a new species the completeness of its autonomy likewise must result from this kind of self-integrity. The formation of a new species is distinct from the simple change that accompanies the growth of the species itself, or, we can say, due to the self-identical completeness of the species, the formation of a new species should not occur haphazardly anywhere at any time. But when we consider that in Australia, where not only kangaroos, but also animals similar to wolves, bears and mice developed from ancestral marsupials, behind the formation of the species there is something else which can be called the self-completeness of a class or of a whole community which directs the evolution of living things. From the point of view of the world of living things, the world system, which was born and developed from one thing, has never been orderless. Just as there was order in the system, there was also order in this development. Development was not the result of chance accumulations, but from the beginning, something like a directedness can be perceived in species' development. Of course, each organism was not conscious of the aim of this directedness. This directedness was the result of the self-completeness inherent not in each living thing, but in the system of this world. Am I finally trapped here in a teleology? Should I put down my pen? Nevertheless, this very self-completeness is the basis of both autonomy and wholeness. It is also the foundation of history and of creation while yet transcending history and creation. All living things and the societies they form through this self-completeness must always be connected with the principle of this world, with the self-completeness of this world.

But what if Australia were joined once again with Asia? The more advanced mammals probably would invade Australia and destroy many of the marsupials that had no refuge. Although Australia is unconnected, this effect has already been seen sporadically as a result of the European invasion together with the various animals they brought. This kind of example is probably about the only occasion when Darwinian natural selection or the survival of the fittest is seen. I want to set this down in the final paragraph of this work, though it concerns me that as a result people may think it contradicts the point of my argument. But ultimately the reorganization of the social structure of the community of living things in Australia always must be the reorganization of the structure based on the world. In order to incorporate Australia, which has until now been a separate world, into one world, a new structural equilibrium is demanded. It is regrettable that the other side of the survival of the fittest is the extinction of the unfit, but we have to think of it from the point of view of the whole world. Those that will become extinct also leave an impression upon the world. In a broad sense, it is also a kind of regeneration from the point of view of the world. But for all that, although I have raised this instance of a clear case of natural selection at the end of this work, people should be aware that natural selection is not, perforce, directly related to the problem of the origin of species.

Original index

bungyō	分業	division of labour; specialization
bunka	文化	culture
bunpu	分布	distribution
chien	地縁	territorial
dōgu	道具	instrument; implement
dōifukugōshakai	同位複合社会	same rank composite society
dōishakai	同位社会	same rank society
eiyō	営養	nutrition
hanshoku	繁殖	breeding; propagation
heikō	平衡	equilibrium; balance
hōkōsei	方向性	direction; bearings
honnō	本能	instinct

hōshin	方針	principle; aim; course
jikan	時間	time; an hour
kaikyū	階級	class
kankyō	環境	environment; surroundings; circumstances
ketsuen (teki)	血縁(的)	blood relation; kin
kinō	機能	faculty; function
kōzō	構造	structure
kūkan	空間	space
ninshiki	認識	cognition; recognition; perception
rekishi	歴史	history
ruien	類縁	family relation; affinity
ruisui	類推	analogical inference; (an) analogy
saibō	細胞	a cell
seikatsukei	生活形	form (pattern) of living (subsistence)
seikatsu no ba	生活の場	field (place) of living

seishin	精神		mind; spirit; soul
sentaku	選択		selection; choice; option
shintai	身体		body; person; (constitution)
shushakai	種社会（種の社会；種の世界）		species society; species world
shutaisei	主体性		subjectivity; independence autonomy; identity
soshitsu	素質		makings; temperament; character
sumiwake	棲み分け		habitat segregation
tekiō	適応		adaptation
tōgōsei	統合性		integration; synthesis; unity
zentaisei	全体性		the whole (nature, attribute)

Bibliography of publications in Western languages by Kinji Imanishi

1930 Mayflies from Japanese Torrents I, *Taiwan hakubutsugaku kaihō* [Taiwan Natural History Bulletin], vol. 30, 263–267.
1932 Mayflies from Japanese Torrents II, *Annot. Zool. Japon*, vol. 8, 525–530.
1933 Mayflies from Japanese Torrents III, *Annot. Zool. Japon*, vol. 13, 64–69.
1934 Mayflies from Japanese Torrents IV, *Annot. Zool. Japon*, vol. 14, 381–395.
1935 Mayflies from Japanese Torrents V, *Annot. Zool. Japon*, vol. 15, 213–221.
1936 Mayflies from Japanese Torrents VI, *Annot. Zool. Japon*, vol. 15, 538–548.
1937 Mayflies from Japanese Torrents VII, *Annot. Zool. Japon*, vol. 16, 321–329, and Mayflies from Japanese Torrents VIII, ibid., 330–337.
1938 Mayflies from Japanese Torrents IX, *Annot. Zool. Japon*, vol. 17, 23–36.
1939 On the altitudinal regions of the northern Japanese Alps, *Bulletin of the Biogeographical Society of Japan*, vol. 9, no. 7, July, 133–144.
1941 Mayflies from Japanese Torrents X, *Memoirs of the College of Science, Kyoto Imp. Univ., Ser. B*, vol. 16, 1–35.
1950 Ecological observations on the Great Khingan Expedition, *Geographical Review*, vol. 40, 236–253.
1950 *Misaki-uma no shaka chōsa: Hōkoku dai 3- ima made ni kokoromita chōsa no yōyaku* (Social life of semi-wild horses in Toimisaki III: Summary for the three surveys undertaken in 1948–49. *Seiri Seitai*, IV [Physiology and Ecology] (in Japanese with English summary), 28–41.
1951 *Naimōko sōgen no ruikeizuke – 1944– nen made ni erareta chishiki no seiri* (Types of vegetation in Inner Mongolia based on the reports of surveys up to 1944), *Shizen to Bunka* [Nature and Culture], Kyoto II (in Japanese with English summary).
1952 *Naimōko sōgen no chiriteki – toku ni keisōgen o chūshin toshite* (Geographical distribution of grasslands in Inner Mongolia–especially the central steppe), *Yūboku minzoku no shakai to bunka* [Nomadic Peoples' Society and Culture], Kyoto (in Japanese with English summary), 129–175.
1954 Nomadism, an ecological interpretation. *Silver Jubilee Vol. of the Jinbun kagaku kenkyūsho* [Institute for Humanistic Studies], Kyoto University, 466–479.
1954 Annapurna and Manaslu, *Himalayan Journal*, 18, 176–177.

Bibliography 91

1955 Scientific activities of 1952 and 1953 expeditions. In Hitoshi Kihara (ed.) *Fauna and Flora of Nepal Himalaya*. Fauna and Flora Research Society; Scientific Results of the Japanese Expeditions to Nepal Himalaya 1952–1953, Kyoto, 1–3.

1957 Social Behavior in Japanese Monkeys, *Macaca fuscata, Psychologia*, vol. 1, 47–54.

1957 Conservation of Japanese Monkeys. *Proceedings and Papers of Sixth Technical Meeting, International Union for the Conservation of Nature and Natural Resources*, 71–72.

1958 Gorillas: A preliminary survey in 1958, *Primates*, vol. 1, 73–78.

1960 Social organization of subhuman primates in their natural habitat, *Current Anthropology*, vol. 1 (5–6), 393–407.

1961 The origin of human family: A primatological approach, *The Japanese Journal of Ethnology 35/3*, 119–138 (In Japanese with English summary). Reprinted in K. Imanishi and Stuart A. Altmann (eds) 1965, *Japanese Monkeys. A Collection of Translations*. Edmonton, 113–140.

1963 Social behavior in Japanese monkeys, *Macaca fuscata*. In Charles H. Southwick (ed.), *Primate Social Behavior*. New Jersey: D van Nostrand Co. Ltd, 68–81.

1963 Editor, *Personality and Health in Hunza Valley*. Results of the Kyoto University Scientific Expedition to the Karakoram and Hindukush, 1955, vol. V., Kyoto University,[Preface by Imanishi, pp. i–iv].

1964 The evolution of personality, *Zinbun* [=*Jinbun*], Memoire of the Research Institute for Humanistic Studies, vol. 7 (Results of the Kyoto University Scientific Expedition to the Karakoram and Hindu Kush), 1–12.

1964 The individual in the society of Japanese monkeys, *Japan Quarterly*, vol. 11, 293–300.

1965 Identification: A process of socialization in the subhuman society of *Macaca fuscata*. In K. Imanishi and S. A. Altmann (eds) *Japanese Monkeys. A Collection of Translations*. Edmonton, 30–51.

1966 The purpose and method of our research in Africa. *Kyoto University African Studies*, vol. 1 (ed.) K. Imanishi, 1–10.

1970 Field studies on primate societies. Twenty years of Japanese research and prospects for the future. In Hideki Yukawa (ed.) *Profiles of Japanese Science and Scientists*. Tokyo: Kodansha Ltd, 30–42.

1972 L'influence qu'exerce l'environnement social du stade immature sur la détermination de la hiérarchie chez les singes japonais. *Colloques Internationaux du Centre national de la recherche scientifique*, no. 198, 149–154.

1984 A proposal for shizengaku: The conclusion to my study of evolutionary theory. *Journal of Social and Biological Structures*, 7, 357–368.

Index

Academic Alpine Club of Kyoto (AACK) xxx
action; function vs. 18; living and 68; recognition and 31
adaptation 76, 78, 80
affinity xxxix, 4–6, 29, 51, 53, 81; differences and 38, 49; discontinuity of 60; food and 29; between predators and prey 56–7; recognition of 5; societal complexes and 58; between species 51; synusiae and 49, 55
aggregation 15, 41, 63
ammonites 81, 82
amoebae 4, 5, 16; environment and 25
amphibians 69
analogy 5n57
animals *see also* mammals; aquatic 51; as automata 6, 7; diversification in plants vs. 58; division into synusiae 50–1; domestication of 73; ecology 57–8; fertilization in 35; forms of 11; lower *see* lower animals; reaction to humans 6; recognition of 10; societal levels 58–9; taxonomy 60; terrestrial 51–2; variation among 73
anthropology xx
anthropomorphism 6, 7, 71, 72
anti-Darwinianism xx–xxi, xxxvii
anti-science xxxvii
anti-selectionism xxi n24
 see also natural selection
ants 44–5
aquatic animals *see* marine animals
artificial selection 73–4

atoms 18, 21–2
Australia; flora and fauna of 84–6
automata 6, 7, 22, 30, 34
autonomy xxxvi, 31, 32, 34, 35–6, 74; of ants 44; division of 82–3; modulation of variability and 79, 82; over environment 39, 42, 44, 58, 62, 64, 70, 75; of physical characteristics 35–6; subjective character and xli; wholeness and 62–3, 71, 85, 86

bacteria 71
balance of forces 37, 41, 52
beauty 81
Bester, John xlii–xliii
biogeographers 57
biology xl, 7, 22, 29
biosociology xx
breeding, interspecific 83
butterflies 12, 45–6

carnivores 58–9
Carpenter, Clarence Ray xlix
caterpillars 45
cells 12, 14, 15, 19, 23; continuous creation of 21; living things and 16, 21; maintenance of selves 24; organic integrated bodies and 23; as organisms 15
Cenozoic era 67
change 18
chemistry 22; organic vs. inorganic 16
chikaku 3n54 *see also* perception
China xxxi, xxxii

Index 93

climate; change in 47–8; differences and 82; distinctive features and 83; plant 34
communities; development of 65–6; life form and 53, 58; phylogenetic 54, 58, 65
comparative psychology 6
conception 23
conflict, avoidance of 41, 42
consciousness 29, 30–1, 35; of self xxxv, 40–1
conservatism 42, 71, 79, 80
conspecifics xli; gathering of 46; recognition of 40
content of living 37–8
continents, division of 84
creation; evolution and 68, 85; re-creation and 17, 24; regeneration and 67; of self 17
"creature" 16
cultivation 42
culture 81, 84
cytology 22

Darwin, Charles 72, 73
death 12, 17, 18
determinism, environmental 34, 59, 74
development, of living things 15
difference(s); affinity and 38, 49; morphology and 39; in predispositions 82; recognition of 3; similarity and 1–8, 37, 46; in tendencies 82
differentiation 31; among individuals within species 63; of cells 23; of parts 62
discontinuities 57, 58–9, 60, 63
distribution ranges 40, 63
division of labor 56, 63, 64, 71
dōbutsu 16
dōifukugōshakai 53n64
dōishakai 48n63
dragonflies 69
dying vs. living 74, 75

earth, inequality of surface 47, 49, 56, 59–60, 82
ecology 39, 51, 57
elephants 67
embryology 16–17

enemies *see* predators
environment xl–i, 21–32, 33; adaptation to 78; autonomy over 42, 44, 62, 64, 70, 75; body as extension of 74, 75; content of living and 38; creation of organisms within 70; as extension of living things xli, 27, 38; as inorganic 36; life necessities and 26; living requirements within 38–9; meanings of 25, 46; recognition of xli, 28–30, 74; ways of life and 81
equilibrium 26, 37, 41, 48, 49, 52, 59, 64, 65, 79, 82
evolution; breadth of 54; creation and 68, 85; of mammals 68; rate of 68; ruling class and 73
extinction 69
eyes, use of 9, 35

family, of ants 45
fertilization, in animals 35
field of living 27, 33, 46
fish 12, 13, 54–5, 66, 68
food; chains 55–6; as extension of body 28, 29; ingestion of 25, 26–7, 35; recognition by lower animals 29
forces, balance of 37, 41, 52
forests 43, 50, 51, 57, 58, 59
Fukuzawa, Yukichi xxxiii–iv
function(s); action vs. 18; structure vs. xl, 13–15, 18, 24

gathering; of conspecifics 46; of members of species xli, 38, 39, 41
gazelles 50–1, 57–8
genetic mutation 73
grasshoppers 50–1, 57–8
grassland 50
groups; herbivores in 43; plants in 43
growth, and living 13, 15, 17, 18

habitat segregation 48, 49, 51–2, 52–3, 63–4, 65
habitats 5, 47–8
Halstead, Beverly xxi
herbivores 43–4, 67
heredity 23, 37

hierarchy, in societies
  see society: classes of
holism 14, 60, 61
holospecia xlii
hunter-gatherers xxxi
hunting societies 42

Imanishi, Kinji; anti-Darwinianism
  of writings xx–xxi, xxxvii;
  anti-science writings xxxvii;
  children of xxx; early life xxix;
  education of xxxii; as follower of
  Darwin xxxvii n44; Kitarō Nishida's
  views compared with xxxv–vii;
  at Kyoto University xxxi–ii,
  xxix–xxx; in Mongolia xxxi;
  mountaineering of xxix–xxx, xxxi;
  studies mayfly larvae xxx; study of
  hunter-gatherers and nomadic
  pastoralists xxxi; surveys of great
  ape species xxxi; wartime
  experiences xxxi; writing of *The
  World of Living Things* xxx–i
imitation 41
Imperial Rescript on Education
  (1890) xxxiv
implements, as extension
  of body 28
independence, and science xxxiv
individualism xxxvii
individual(s) 40 *see also* members of
  species; differentiation among 63;
  distribution of 63; maintenance 24,
  36–7; origin of 77; relationship
  among 41; species vs. 62, 63
inequality 47
insects 66, 70; community of 53;
  as parasites 44; as ruling class
  68–9, 70, 71; society of 43;
  synusial complexes 53, 60
instinct 4, 30–1, 31–2, 46, 74
integrity 15, 22–3, 24, 30, 32, 85
intuition 4
isolation 2, 37, 39, 42
Itani, Jun'ichirō xx, xxii, xxxi
Itō, Shuntarō xxxviii

Japan; contact with West xxxii–iv;
  science in xxi n27, xxxiii–iv;
  Western technology and xxxiii–iv

Japanese Enlightenment xxxiii–iv
Japanese macaques xxxi

kangaroos 85
Kani, Tōkichi xxxi
Kasuya, Eiichi xxxi n35
Kawakatsu, Heita xix, xxii, xxxviii
Kawakita, Jiro xxxi
Kawamura, Shunzō xx
killer whales 42, 44
kinship 38, 39, 61, 76
Kira, Tatsuo xxxi
Kyoto School of Philosophy xxxiv

Laramide revolution 66
larvae 45
legume family 58
life; maintenance of 24;
  of matter 21–2; recognition
  of necessities for 26
life energy 37
life field *see* field of living
life form(s) 39, 42, 48, 50, 52,
  53, 54
lifestyle 74, 76, 81
living *see also* field of living;
  acting and 68; dying vs. 74, 75;
  growth and 13
living things *see also* individual(s);
  organisms; cells and 15, 16, 21;
  creation of selves 17, 21; death of
  17; decay of 17; development
  of 15; environment as extension
  of xli, 27, 38; forms of
  10–11, 11–12; function(s) of
  see function(s); growth of 13, 15,
  17, 18; as independent systems 25;
  life of 15–16, 22, 26; maintenance
  of selves 17, 24; nonliving things
  vs. 5, 7, 9, 10–11, 19–20; organic
  integrated bodies and 15, 21,
  23–5, 26; as organisms 15;
  parts vs. whole 23; peaceful
  life and 26; plants as 22; society
  of 46; spatial existence of 17;
  structure of *see* structure;
  wholeness in 23
lower animals; recognition of
  environment 28–30;
  recognition of food 29

Index 95

macroevolution 78
maintenance; of cells 24;
  of individuals 24, 36–7;
  of life 21, 24; of living things 17,
  24, 36–7; of species 36;
  of world 36
mammals 43, 69, 77; ancestors of 77;
  development of 67–8; emergence
  of 68, 69
mankind 71, 72
Marco Polo Bridge Incident li n52
marine animals 59
marsupials 84, 85, 86
matter 20, 21, 33, 36
mayfly larvae xxx
mechanistic thinking xxxvi, 75
Meiji era xxxii
Meirokusha society xxxiii
members of species *see also*
  individual(s); aggregating of 41, 63;
  dispersion of 39–40, 42;
  gathering of xli, 38, 39;
  mutual intolerance 38
Mesozoic era 66–7
metabolism 17, 36
microclimate 34
microevolution 78
microscopic things 10
molecules 21
Mongolia xxxi
monkeys 4, 5, 43 *see also* Japanese
  macaques
monotremes 84
morphology 10; difference and 39;
  life forms and 54
Morris-Suzuki, Tessa xxxviii
mutability 78–9
mutation 82
mutual interactions 6, 37, 38
mutual intolerance 38, 49

Nakao, Sasuke xxxi
natural selection xlii, 73, 75, 77,
  78, 79, 86
nature, study of xxxvii–viii
nihonjinron xxi n27
ninshiki 3n54 *see also* recognition
Nishida, Kitarō xxxiv–vii, xxxviii
nomadic pastoralists xxxi
nomadism 43

nonliving things 18–20; forms of 11;
  living things vs. 5, 7, 9, 10–11,
  19–20; structure and function 19

objectivity xxxv
offspring; similarity to parents 3, 37;
  variation in 76
olfaction *see* smell
Opium War (1840) xxxii–iii
organic integrated bodies 15, 21–2,
  23–5; living things and 26
organisms 14, 16, 22 *see also* living
  things; cells as 15; living things
  as 15
Ōyashima 84

parasites 43–4
parasitism 56
parents; offspring's similarity to 3, 37;
  variation in offspring 76
parts vs. whole 62
pastoralism 42
perception 3
phylogenetic communities 54, 58, 65
physics 22
physiology 36
plants; climate 34; communities of 54;
  diversification in animals vs. 58;
  division into synusiae 49–50, 51;
  ecology 57; environment and 25;
  in groups 43; as living things 22;
  phylogenetic communities
  among 58; pollination
  see pollination; recognition by 35;
  societies of 46, 49–50; within
  species 40
pollination 35, 38
predators 25, 26, 42, 44, 55–7, 59,
  64, 66
predispositions, difference in and
  difference in tendencies 82
prey, predators vs. 55–7, 59, 64, 66
primary synusia 53
primatology xi–xii, xix, xx, xxxi
protoconsciousness 31
protozoa 40

race 40
random mutation xlii
random variation 73, 75, 76

reaction, as recognition 34–5, 74
recognition 3, 29; action and 31; of affinity xxxix, 5, 29; consciousness and 35; of conspecifics 40; of differences 3; of environment xli, 28–30, 74; of food 29; naive 4; of necessities for life 26, 27; of own smell and voice 40–1; by plants 35; reaction as 34–5, 74; subjective response of 6; of unessential things 27
regeneration 67, 86
relationships xxxix, 2, 6; among individuals 41
reproduction 36–7, 38, 41–2, 45, 62, 75 *see also* breeding, interspecific
reptiles 68, 77; age of 69
reptiles, age of 66–7
resemblance *see* similarity
rose family 58
ruien 4n56 *see also* affinity
ruisui 5n57
ruling class 66–7, 68–72, 73
Russo-Japanese War xxxii

Schultz, Adolf xlviii
scientific development, differences in 83
seibutsu xlii, 16
Seihoku Kenkyūsho xxxi
seikatsu 27
seimei 27
seishin 33n61
sei-tai 16
self-consciousness xxxv, 40–1
self-destruction 66
self-substantialization xxxv
shizengaku xxxvii–viii
shokubutsu 16
shrubs 50
shutaisei xli
similarity; difference and 1–8, 37, 46; in variation 76–7
Sino-Japanese war li n52
size; differences 44, 57–8, 58–9; discontinuities 58–9, 60
skills 73; hereditary 70–1; specialization of 70, 80
smell, recognition of 40–1
sociality 46

societal complexes 57–8
society xxxvi–vii, 8; of ants 44–5; classes of 66–7, 69, 70–1 *see also* ruling class; as place of shared living 41; ranks within 70–1, 73; same rank 48; spatial–structural aspect 46–7; species and 46; survival within 41
solar system 18
space, and time 17–19, 18–19, 23–4, 41, 61
specia xxxvi, xli, 48n63
specialization; differentiation vs. 23, 31; of skills 70, 80
species; affinity between 51; cultural characteristics of 80–2, 83; death of 77; destiny of 85; differentiation among individuals within 63; distribution 48, 63; formation of new 67, 77; forming synusial complexes 54; genetic predisposition of 82; independence of 83; independent integrity 85; individuals and 62, 63; individuals within *see* members of species; maintenance of 36; migration of 84; origin of 72, 76–7, 86; plants within 40; separation of 58–9, 82–3; society and 46; survival within 41
species society xxxvi–vii, xli, 61, 64, 67; spatiality of 62
sperm 35
structure xl, 2, 9–20, 12 *see also* morphology; function vs. 13–15, 18, 24; of living vs. nonliving things 12
subjectivity xxxv, 5–6, 7
subsistence 42, 81
survival; adaptation and 78; characteristics nonessential for 80, 83; competition for xxxvii n44, 38; of fittest 75, 79, 86; society and 41; struggle for 76
sustenance 42, 47
synusia(e) 48–55, 64, 67, 69, 83; affinity and 55; plant vs. animal 54; societal complexes and 55; species societies and 61
synusial complexes 53–9, 61, 64–5, 67, 69; division of 59, 61; insects and 60; phylogenetic 54, 58, 61;

predator/prey division 59;
  size differences and 58–9;
  societal complexes and 55
synusial species 82

Takasaki, Hiroyuki xx, xxii
Tanzania xxxi
tape worms 52
taxonomy xl, 7, 10, 29, 51, 60,
  81–2
tendencies, difference in and difference
  in predispositions 82
terrestrial animals 59
territorial relations 39
territorial relationships 61
territoriality 48
tetsugaku xxxiv
things; categorization of 9; living vs.
  nonliving 9–10; microscopic 10;
  perception of 9; similarity, and
  heredity 37
time, and space 17–19, 18–19,
  23–4, 41, 61

Tokugawa era xxxii, xxxviii
trees, society of 50, 54 *see also* forests

Ueyama, Syunpei xxxi n33, xxxv,
  xxxviii
ultimate society 61–2
Umesao, Tadao xxxi
ungulates 43, 44

variability, modulation of 79
variation 75–8, 79–80; random
  see random variation
vertebrates 69, 71
vision 40
vocalization, recognition of 40–1

Washburn, Sherwood xlix
whales 43, 54–5
whole vs. parts 62
wholeness 22, 23, 32, 65;
  autonomy and 62–3, 71

Yagi, Shusuke xxii